SURFACTANTS IN AGROCHEMICALS

SURFACTANT SCIENCE SERIES

1. Nonionic Surfactants, *edited by Martin J. Schick* (see also Volumes 19 and 23)
2. Solvent Properties of Surfactant Solutions, *edited by Kozo Shinoda* (out of print)
3. Surfactant Biodegradation, *R. D. Swisher* (see Volume 18)
4. Cationic Surfactants, *edited by Eric Jungermann* (see also Volumes 34, 37, and 53)
5. Detergency: Theory and Test Methods (in three parts), *edited by W. G. Cutler and R. C. Davis* (see also Volume 20)
6. Emulsions and Emulsion Technology (in three parts), *edited by Kenneth J. Lissant*
7. Anionic Surfactants (in two parts), *edited by Warner M. Linfield* (out of print)
8. Anionic Surfactants: Chemical Analysis, *edited by John Cross* (out of print)
9. Stabilization of Colloidal Dispersions by Polymer Adsorption, *Tatsuo Sato and Richard Ruch*
10. Anionic Surfactants: Biochemistry, Toxicology, Dermatology, *edited by Christian Gloxhuber* (see Volume 43)
11. Anionic Surfactants: Physical Chemistry of Surfactant Action, *edited by E. H. Lucassen-Reynders* (out of print)
12. Amphoteric Surfactants, *edited by B. R. Bluestein and Clifford L. Hilton* (out of print)
13. Demulsification: Industrial Applications, *Kenneth J. Lissant*
14. Surfactants in Textile Processing, *Arved Datyner*
15. Electrical Phenomena at Interfaces: Fundamentals, Measurements, and Applications, *edited by Ayao Kitahara and Akira Watanabe*
16. Surfactants in Cosmetics, *edited by Martin M. Rieger*
17. Interfacial Phenomena: Equilibrium and Dynamic Effects, *Clarence A. Miller and P. Neogi*
18. Surfactant Biodegradation: Second Edition, Revised and Expanded, *R. D. Swisher*

19. Nonionic Surfactants: Chemical Analysis, *edited by John Cross*
20. Detergency: Theory and Technology, *edited by W. Gale Cutler and Erik Kissa*
21. Interfacial Phenomena in Apolar Media, *edited by Hans-Friedrich Eicke and Geoffrey D. Parfitt*
22. Surfactant Solutions: New Methods of Investigation, *edited by Raoul Zana*
23. Nonionic Surfactants: Physical Chemistry, *edited by Martin J. Schick*
24. Microemulsion Systems, *edited by Henri L. Rosano and Marc Clausse*
25. Biosurfactants and Biotechnology, *edited by Naim Kosaric, W. L. Cairns, and Neil C. C. Gray*
26. Surfactants in Emerging Technologies, *edited by Milton J. Rosen*
27. Reagents in Mineral Technology, *edited by P. Somasundaran and Brij M. Moudgil*
28. Surfactants in Chemical/Process Engineering, *edited by Darsh T. Wasan, Martin E. Ginn, and Dinesh O. Shah*
29. Thin Liquid Films, *edited by I. B. Ivanov*
30. Microemulsions and Related Systems: Formulation, Solvency, and Physical Properties, *edited by Maurice Bourrel and Robert S. Schecter*
31. Crystallization and Polymorphism of Fats and Fatty Acids, *edited by Nissim Garti and Kiyotaka Sato*
32. Interfacial Phenomena in Coal Technology, *edited by Gregory D. Botsaris and Yuli M. Glazman*
33. Surfactant-Based Separation Processes, *edited by John F. Scamehorn and Jeffrey H. Harwell*
34. Cationic Surfactants: Organic Chemistry, *edited by James M. Richmond*
35. Alkylene Oxides and Their Polymers, *F. E. Bailey, Jr., and Joseph V. Koleske*
36. Interfacial Phenomena in Petroleum Recovery, *edited by Norman R. Morrow*
37. Cationic Surfactants: Physical Chemistry, *edited by Donn N. Rubingh and Paul M. Holland*
38. Kinetics and Catalysis in Microheterogeneous Systems, *edited by M. Grätzel and K. Kalyanasundaram*
39. Interfacial Phenomena in Biological Systems, *edited by Max Bender*
40. Analysis of Surfactants, *Thomas M. Schmitt*
41. Light Scattering by Liquid Surfaces and Complementary Techniques, *edited by Dominique Langevin*
42. Polymeric Surfactants, *Irja Piirma*
43. Anionic Surfactants: Biochemistry, Toxicology, Dermatology. Second Edition, Revised and Expanded, *edited by Christian Gloxhuber and Klaus Künstler*
44. Organized Solutions: Surfactants in Science and Technology, *edited by Stig E. Friberg and Björn Lindman*
45. Defoaming: Theory and Industrial Applications, *edited by P. R. Garrett*

46. Mixed Surfactant Systems, *edited by Keizo Ogino and Masahiko Abe*
47. Coagulation and Flocculation: Theory and Applications, *edited by Bohuslav Dobiáš*
48. Biosurfactants: Production · Properties · Applications, *edited by Naim Kosaric*
49. Wettability, *edited by John C. Berg*
50. Fluorinated Surfactants: Synthesis · Properties · Applications, *Erik Kissa*
51. Surface and Colloid Chemistry in Advanced Ceramics Processing, *edited by Robert J. Pugh and Lennart Bergström*
52. Technological Applications of Dispersions, *edited by Robert B. McKay*
53. Cationic Surfactants: Analytical and Biological Evaluation, *edited by John Cross and Edward J. Singer*
54. Surfactants in Agrochemicals, *Tharwat F. Tadros*

ADDITIONAL VOLUMES IN PREPARATION

Solubilization in Surfactant Aggregates, *edited by Sherril D. Christian and John F. Scamehorn*

SURFACTANTS IN AGROCHEMICALS

Tharwat F. Tadros
Zeneca Agrochemicals
Berkshire, England

Marcel Dekker, Inc. New York • Basel • Hong Kong

Library of Congress Cataloging-in-Publication Data

Tadros, Th. F.
Surfactants in agrochemicals / Tharwat F. Tadros.
p. cm. – (Surfactant science series ; 54)
Includes bibliographical references and index.
ISBN 0-8247-9100-2 (hardcover)
1. Surface active agents. 2. Agricultural chemicals–Adjuvants. I. Title. II. Series: Surfactant science series ; v. 54.
S587.73.S95T33 1994
668'.6–dc20 94-37159
CIP

The publisher offers special discounts on this book when ordered in bulk quantities. For more information, write to Special Sales/Professional Marketing at the address below.

This book is printed on acid-free paper.

Marcel Dekker, Inc.
270 Madison Ave, New York, New York, 10016

Current printing (last digit):

10 9 8 7 6 5 4 3 2 1

PRINTED IN THE UNITED STATES OF AMERICA

Preface

Surfactants play a major role in agrochemicals in both formulation and optimization of biological efficacy. To formulate any agrochemical as a dispersion or dispersible system that is suitable and economic for application, one needs to use a suitable surfactant. Surfactants are essential for the preparation of any disperse system, such as an oil/water emulsion, a suspension concentrate, and a microemulsion. Surfactants are also essential for maintaining the long-term physical stability of these systems and their application as sprays. Dispersible systems such as wettable powders, emulsifiable concentrates, and water-dispersible grains also require surfactants for their application. In all formulations, the surfactant plays a crucial role in enhancing and optimizing the biological efficacy of agrochemicals. Indeed, with many agrochemicals, a high concentration of surfactant may be required for ensuring the efficacy of the chemical.

In spite of the above obvious roles of surfactants in agrochemicals, fundamental studies of their role in formulation preparation, and optimization of biological efficacy are, to date, far from satisfactory.

Most formulation chemists choose their surfactants using trial and error. As a result, each time a new agrochemical is discovered the formulation chemist repeats this tedious procedure to find the most suitable surfactant. This is particularly the case for discovering the surfactant that gives optimal biological efficacy.

This book has been written with the objective of introducing a fundamental approach to the selection of surfactants in agrochemical formulations. After a short general introduction (Chapter 1) to illustrate the role of surfactants in various agrochemicals formulations, Chapter 2 is devoted to the physical chemistry of surfactant solutions. This fundamental chapter summarizes the unusual properties of surfactant solutions, in particular their micellization and the formation of structural units as spherical, rod shaped, and lamellar systems, which play a major role in biological efficacy. Chapter 3 deals with the adsorption of surfactants and polymers (sometimes referred to as macromolecular surfactants) at the air/liquid, liquid/liquid, and solid/liquid interfaces. This subject is essential for understanding how surfactants and polymers can stabilize disperse systems such as emulsions, suspensions and microemulsions.

Chapter 4 deals with the subject of emulsifiable concentrates, which lack fundamental understanding for their formulation. An example is given to illustrate how a fundamental approach may be used to understand the formation of these systems and their spontaneity of emulsification on dilution. Chapter 5 presents a great deal of fundamental knowledge on the preparation and subsequent stabilization of emulsions. Suspension concentrates are similarly described in Chapter 6. A potentially useful type of formulation is a microemulsion which to date has seldom been used in the agrochemical industry. The advantages of microemulsions over emulsions are their inherent thermodynamic stability and their potential to enhance biological efficacy; this topic is covered in some detail in Chapter 7. The last chapter discusses the role of surfactants in transfer and performance of agrochemicals. This is perhaps the most difficult subject to deal with at a fundamental level. However, an attempt has been made to describe the role of surfactants in spray droplet formation and in wetting and spreading, as well as their impaction adhesion and retention. The role of surfactants in deposit formation, solubilization, and their effect on transport is briefly described to illustrate how surfactants may enhance the biological efficacy of an agrochemical.

It should be mentioned that this text is far from exhaustive. I have concentrated on topics on which I have carried out some fundamental research.

Since this book was written at home in my spare time, in particular during Christmas vacations, I am indebted to my wife and children for their tolerance of my spending the time writing rather than enjoying various activities with them.

Tharwat F. Tadros

Contents

SURFACTANTS IN AGROCHEMICALS

1

General Introduction

The formulations of agrochemicals cover a wide range of systems that are prepared to suit a specific application. In some cases, an agrochemical is a water-soluble compound of which paraquat and glyphosate (both herbicides) are probably the most familiar. Paraquat is a 2,2 bypyridium salt and the counterions are normally chloride. It is formulated as a 20% aqueous solution that on application is simply diluted into water at various ratios (1:50 up to 1:200 depending on the application). To such an aqueous solution, surface active agents (sometimes referred to as wetters) are added that are essential for a number of reasons. The most obvious reason for adding surfactants is to enable the spray solution to adhere to the target surface and spread over a large area. However, such a picture is an oversimplification since the surface active agent plays a more subtle role in the optimization of the biological efficacy. Thus, the choice of the surfactant system in an agrochemical formulation is crucial since it has to perform a number of functions. To date, such a choice is made by a trial and error procedure, due to the complex nature of application and lack of understanding of the mode of action of the chemical. It is the objective of this book to rationalize the role of surfactants in formulations, subsequent application, and optimization of biological efficacy.

Most agrochemicals are water insoluble compounds with various physical properties, which have first to be determined in order to decide on the type of formulation. One of the earliest types of formulations are wettable powders (WPs) that are suitable for formulating solid water insoluble compounds that can be produced in a powder form. The chemical (which may be micronized) is mixed with a filler such as china clay and a solid surfactant such as sodium alkyl or alkylaryl sulfate or sulfonate is added. When the powder is added to water, the particles are spontaneously wetted by the medium and on agitation dispersion of the particles takes place. It is clear that the particles should remain suspended in the continuous medium for a period of time depending on the application. Some physical testing methods are available to evaluate the suspensibility of the WP. Clearly the surfactant system plays a crucial role in wettable powders. In the first place it enables spontaneous wetting and dispersion of the particles. Also, by adsorption on the particle surface, it provides a repulsive force that prevents aggregation of the particles. Such a process of aggregation will enhance the settling of the particles and may also cause problems on application such as nozzle blockage.

The second and most familiar type of agrochemical formulations is the emulsifiable concentrates (ECs). This is produced by mixing an agrochemical oil with another one such as xylene or trimethylbenzene or a mixture of various hydrocarbon solvents. Alternatively, a solid pesticide could be dissolved in a specific oil to produce a concentrated solution. In some cases, the pesticide oil may be used without any extra addition of oils. In all cases, a surfactant system (usually a mixture of two or three components) is added for a number of purposes. First, the surfactant enables self-emulsification of the oil on addition to water. This occurs by a complex mechanism that involves a number of physical changes such as lowering of the interfacial tension at the oil/water interface and enhancement of turbulence at that interface with the result of spontaneous production of droplets.

Second, the surfactant film that adsorbs at the oil/water interface stabilizes the produced emulsion against flocculation and/or coalescence. As we will see in later sections, emulsion breakdown must be prevented, otherwise excessive creaming or sedimentation or oil separation may take place during application. This results in an inhomogeneous application of the agrochemical on the one hand and possible losses on the other. The third role of the surfactant system in agrochemicals is in enhancement of biological efficacy. As we will see in later chapters, it is essential to arrive at optimum conditions for effective use of the agrochemicals. In this case, the surfactant system will help spreading of the pesticide at the target surface and may enhance its penetration.

In recent years, there has been great demand to replace ECs with concentrated aqueous oil-in-water (o/w) emulsions. Several advantages may be envisaged for such replacements. In the first place, one is able to replace the added oil with water, which is much cheaper and environmentally acceptable. Second, removal of the oil could help in reducing undesirable effects such as phytotoxicity, skin irritation, etc. Third, by formulating the pesticide as an o/w emulsion, the droplet size can be kept to an optimum value, which may be crucial for biological efficacy. Fourth, water soluble surfactants, which may be desirable for biological optimization, can be added to the aqueous continuous phase. As we will see in later chapters, the choice of a surfactant or a mixed surfactant system is crucial for preparation of a stable o/w emulsion. In recent years, macromolecular surfactants have been designed to produce very stable o/w emulsions that could be easily diluted into water and applied without any detrimental effects to the emulsion droplets.

A similar concept has been applied to replace WPs, namely, with aqueous suspension concentrates (SCs). These systems are more familiar than EWs and they have been introduced for several decades. Indeed, SCs are probably the most widely used systems in agrochemical formulations. Again, SCs are much more convenient to apply than WPs. Dust hazards are absent, and the formulation can be simply diluted in the spray tanks, without the need of any vigorous agitation. As we will see later, SCs are produced by a two- or three-stage process. The pesticide powder is first dispersed in an aqueous solution of a surfactant or a macromolecule (usually referred to as the dispersing agent) using a high-speed mixer. The resulting suspension is then subjected to a wet-milling process (usually bead milling) to break any remaining aggregates or agglomerates and reduce the particle size to smaller values. One usually aims at a particle size distribution ranging from 0.1 to 5 μm, with an average of 1–2 μm. The surfactant or polymer added adsorbs on the particle surfaces and resulting in their colloidal stability. The particles need to be maintained stable over a long period of time, since any strong aggregation in the system may cause various problems. First, the aggregates being larger than the primary particles tend to settle faster. Second, any gross aggregation may result in lack of dispersion on dilution. The large aggregates can block the spray nozzles and may reduce biological efficacy as a result of the inhomogeneous distribution of the particles on the target surface. Apart from their role in ensuring the colloidal stability of the suspension, surfactants are added to many SCs to enhance their biological efficacy. This is usually produced by solubilization of the insoluble compound in the surfactant micelles. This will be discussed in later chapters. Another role that a surfactant may play in SCs is the reduction of crystal growth (Ostwald ripening). The latter process may occur when the solubility of the agrochemical is appreciable (say, greater than 100 ppm) and when the SC is polydisperse. The smaller particles will have higher solubility than the larger ones. With time, the small particles dissolve and become deposited on the larger one. Surfactants may reduce this Ostwald ripening by adsorption on the crystal surfaces, thus preventing deposition of the molecules at the surface. This will be described in detail in the chapter on SCs.

Very recently, microemulsions are being considered as potential systems for formulating agrochemicals. Microemulsions are isotropic, thermodynamically stable systems consisting of oil, water, and surfactant(s) whereby the free energy of formation of the system is zero or negative. It is obvious as to why such systems, if formulatable, are very attractive since they will have an indefinite shelf life (within a certain temperature range). Since the droplet size of microemulsions is very small (usually less than 50 nm), they appear transparent. As we will see in later chapters, the microemulsion droplets may be considered as swollen micelles and hence they will solubilize the agrochemical. This may result in considerable enhancement of the biological efficacy. Thus, microemulsions may offer several advantages over the commonly used macroemulsions. Unfortunately, formulating the agrochemical as a microemulsion is not straightforward since one usually uses two or more surfactants, an oil and the agrochemical. These tertiary systems produce various complex phases and it is essential to investigate the phase diagram before arriving at the optimum composition of microemulsion formulation. As we will see in the chapter on microemulsions, a high concentration of surfactant (10–20%) is needed to produce such formulation. This makes such systems more expensive to produce than macroemulsions. However, the extra cost incurred could be offset by an enhancement of biological efficacy which means that a lower agrochemical application rate could be achieved.

The above general introduction, illustrates the role of surfactants in various agrochemical formulations. Before dealing with the role of surfactant in each type of these formulations, two basic chapters will be introduced. Chapter 2 will deal with the physical chemistry of surfactant solutions. This includes topics such as properties of surfactant solutions, their micellization, and solubility. The thermodynamics of micelle formulation will also be discussed to illuminate the driving force for providing such aggregation units. Some sections will also be devoted to the properties of concentrated surfactant solutions. As we will see in Chapter 2, at relatively high concentration, surfactants form liquid crystalline phases in aqueous solution that may have a big influence on the performance of the agrochemical. The solubilization process produced by micelles and liquid crystalline units will also be discussed.

Chapter 3 will deal with the process of adsorption of surfactants (and polymers since these are considered as macromolecular surfactants). Understanding adsorption and conformation of surfactants and polymers at the oil/water and solid/liquid interface is crucial in determining the stability/instability of emulsions and suspensions. Some thermodynamic treatments of surfactant adsorption will be described and theories on macromolecular surfactant adsorption will be summarized.

Chapters 4–7 will deal with the role of surfactants in emulsifiable concentrates, emulsions, suspension concentrates, and microemulsions. These chapters will deal with the various fundamental aspects of stability/instability of these systems. A section will be devoted in Chapter 6 to the flow behavior (rheology) of suspensions. This topic is of vital importance in determining the physical characteristics and long term stability of such systems. The chapter on microemulsion will summarize the theories of microemulsion formation and stability, the formulation of such systems, and the methods for their characterization.

Chapter 8 will deal with the role of surfactants in transfer and performance of pesticides. First, the role of surfactants and polymers in spray formation as well as their impaction and adhesion will be considered. This will be followed by a section on droplet sliding and spray retention. The third section will deal with the process of wetting and spreading that are crucial in distribution of the chemical on the target surface. Evaporation of spray drops and deposit formation will be considered. Under these conditions, the concentration of the surfactant in such deposits will increase to levels whereby liquid crystalline units will be formed. Their role in transfer of the pesticide will be considered. The general topic of solubilization and its effect on transport will be considered in Chapter 8. As we will see, solubilization may lead to enhancement of biological efficacy by increasing the flux of the chemical (by diffusion and convection) as a result of its presence in a micellar unit. The role of deposit formation and its effect on tenacity of the agrochemical particles will be briefly discussed.

2

Physical Chemistry of Surfactant Solutions

I. GENERAL CLASSIFICATION OF SURFACE ACTIVE AGENTS

Surface active agents (usually referred to as surfactants) are amphipathic molecules consisting of a nonpolar hydrophobic portion, usually a straight or branched hydrocarbon or fluorocarbon chain containing 8–18 carbon atoms, which is attached to a polar or ionic portion (hydrophilic). The hydrophilic portion can therefore be nonionic, ionic, or zwitterionic accompanied by counterions in the last two cases. The most common hydrophilic nonionic group is that based on ethylene oxide and a general structure for these nonionic surfactants is $C_nH_{2n+1}(CH_2{-}CH_2{-}O)_n{-}OH$ or $C_nH_{2n+1}{-}[C_6H_5{-}(CH_2{-}CH_2O)_n]OH$ (alcohol or alkylphenol ethoxylates, respectively). The most common ionic surfactants are those containing a sulfate or sulfonate head group, i.e., $C_nH_{2n+1}SO_4Na$, $C_nH_{2n+1}SO_3Na$, $C_nH_{2n+1}{-}C_6{-}H_5{-}SO_3Na$ (alkyl sulfates, alkylsulfonates and alkylbenzene sulfonates, respectively). Cationic surfactants are exempli-

Table 2.1 Classification of Surfactants and Some Examples

---

1. **Ionic surfactants**
   - 1.1. Anionic surfactants
     - Sodium dodecyl sulfate (SDS) $C_2H_{25}SO_4Na$
     - Sodium dodecylbenzenesulfonate $C_{12}H_{25}C_6H_5{-}SO_3Na$
   - 1.2. Cationic surfactants
     - Cetyltrimethylammonium chloride $C_{16}H_{33}(CH_3)_3NCl$
     - Didodecyl dimethylammonium chloride

       $C_{12}H_{25}$

       $\diagdown N(CH_3)_2{-}Cl$

       $C_{12}H_{25}$ $\diagup$

2. **Nonionic surfactants**
   - 2.1 Large hydrophilic head group (based on ethylene oxide
     - Dodecyl hexaethylene glycol monoether

       $C_{12}H_{25}(CH_2{-}CH_2{-}O)_6{-}OH$ abbreviated $C_{12}E_6$
     - Nonyl phenol ethoxylates $C_9H_{19}{-}C_6H_5{-}(CH_2{-}CH_2{-}O)_9OH; C_9\varphi E_9$
   - 2.2. Small hydrophilic head group
     - Dodecyl dimethylamine oxide

       CH_3

       |

       $C_{12}H_{25}{-}N{\rightarrow}O$

       |

       CH_3
     - Dodecyl sulphinyl ethanol $C_{12}H_{25}{-}SO{-}CH_2{-}CH_2{-}OH$

3. **Zwitterionic surfactants** (both anionic and cationic groups), e.g., 3-dimethyl dodecyl propane sulphonate (betaines)

   $C_{12}H_{25}{-}N{-}CH_2{-}CH_2{-}CH_2{-}SO_3^-$

   |

   $(CH_3)_2$

   Lecithin, swelling surfactant—a triglyceride

   $C_{17}H_{35}{-}COO{-}CH_2$

   |

   $C_{17}H_{35}{-}COO{-}CH$ O^-

   | |

   $CH_2{-}O{-}P{-}O{-}CH_2{-}CH_2{-}CH_2{-}N^+(CH_3)_3$

   $\downarrow$

   O

Table 2.1 (continued)

---

4. Polymeric surfactants

Partially hydrolysed poly(vinyl acetate) commonly referred to as poly(vinyl alcohol)

$$(CH_2-\underset{\displaystyle OH}{\underset{|}{CH}})_x-(CH_2-\underset{\displaystyle OCOCH_3}{\underset{|}{CH}})_y-(CH_2-\underset{\displaystyle OH}{\underset{|}{CH}})_x-$$

Polyethylene oxide–polypropylene oxide poly(ethylene oxide) block copolymers commonly referrred to as luronics

$$HO-(CH_2-CH_2-O)_x-(CH_2-\underset{\displaystyle CH_3}{\underset{|}{CH}}-O)_y-(CH_2-CH_2-O)_x-OH$$

5. Polyelectrolytes

Naphthalene formaldehyde sulfonated condensates

Lignosulfonates: anionic polyelectrolytes prepared by sulfonation of wood lignin (complex structures)

Polyacrylic acid and polyacrylic/polymethacrylic acid

$$-(CH_2-\underset{\displaystyle COOH}{\underset{|}{CH}}-)_x- \qquad -(CH_2-\underset{\displaystyle COOH}{\underset{|}{CH}}-)_x-(CH-\overset{\displaystyle CH_3}{\overset{|}{\underset{\displaystyle COOH}{\underset{|}{C}}}}-)_y-(CH_2-\underset{\displaystyle COOH}{\underset{|}{CH}})_x-$$

Polyvinylpyridinium salts

---

fied by the alkyltrimethylammonium type, e.g., $C_nH_{2n+1}(CH_3)_3N^+Cl^-$. Zwitterionic surfactants contain both anionic and cationic head groups, e.g., betaines. A simple classification of surfactants, based on the nature of the head group, is given in Table 2.1. A useful technical reference that lists commerical surfactants is *McCutcheon's Emulsifiers and Detergents* [1] which is published annually to update the available list.

As mentioned above, surfactants are amphiphilic substances that have their hydrophilic and hydrophobic portions well separated. The hydrophobic part, i.e., the alkyl or alkylaryl group, interacts weakly with the water molecules in an aqueous environment, whereas the polar or ionic head group interacts strongly with water molecules via dipole–dipole or ion–dipole interactions. It is this strong interaction with the water molecules that renders the molecule soluble in water. However, to avoid the entropically unfavorable contact between water and the hydrophobic portion, the latter is squeezed out of water by the cooperative action of dispersion and hydrogen bonding between the water molecules. This leads to accumulation of the molecules (i.e., their adsorption) at interfaces. In this manner the hydrophobic group will associate with the hydrophobic surface, e.g., solid particles or oil droplets, while leaving the hydrophilic group in contact with water. Self-association into structures with marked hydrophobic–hydrophile separation is another way of achieving this. As we will see in the next section, surfactant self-association is strongly cooperative and starts generally with the formation of roughly spherical units, referred to as micelles, around a fairly well-defined concentration, the critical micelle concentration.

II. PROPERTIES OF SOLUTIONS OF SURFACE ACTIVE AGENTS

The physical properties of surface active agents differ from those of smaller or nonamphipathic molecules in one major aspect, namely, the abrupt changes in their properties above a critical concentration [2]. This is illustrated in Figure 2.1 in which a number of physical properties (osmotic pressure, turbidity, solubilization, magnetic resonance, surface tension, equivalent conductivity, and self-diffusion) are plotted as a function of concentration [3]. All of these properties (interfacial and bulk) show an abrupt change of a particular concentration that is consistent with the fact that at and above this concentration surface active ions or molecules in solution associate to form larger units. These associated units are called micelles and the concentration at which this association phenomenon occurs is known as the critical micelle concentration (cmc).

Each surfactant molecule has a characteristic cmc value at a given temperature and electrolyte concentration. A compilation of cmc val-

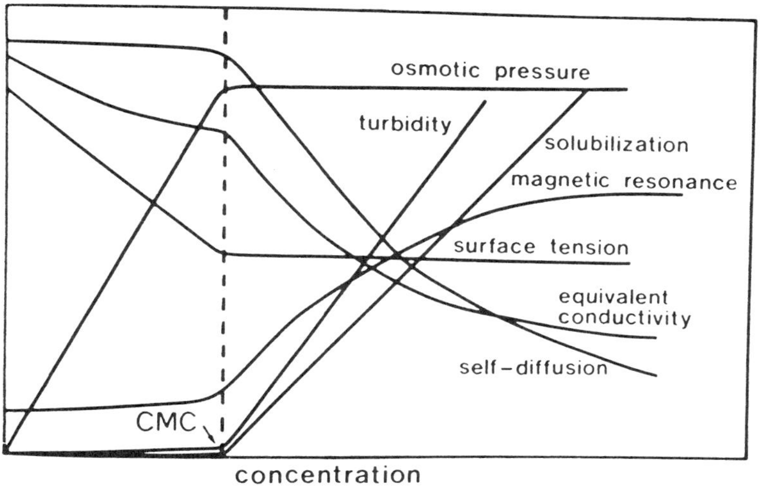

Figure 2.1 Changes in the concentration dependence of a wide range of physicochemical changes around the cmc. After Lindman et al. [3].

ues has been given in 1971 by Mukerjee and Mysels [4], which is clearly not an up-to-date text but is an extremely valuable reference. As an illustration, the cmc values of a number of surface active agents are given in Table 2.2, to show some of the general trends [2]. Within any class of surface active agent, the cmc decreases with an increase in chain length of the hydrophobic portion (alkyl group). With nonionic surfactants, increasing the length of the hydrophilic group (polyethylene oxide) causes an increase in cmc. In general, nonionic surfactants have lower cmc values than their corresponding ionic surfactants of the same alkyl chain length. Incorporation of a phenyl group in the alkyl group increases its hydrophobicity to a much smaller extent than increasing its chain length with the same number of carbon atoms.

The presence of micelles can account for many of the unusual properties of solutions of surface active agents. For example, it can account for the near constant surface tension value, above the cmc (see Fig. 2.1). It also accounts for the reduction in molar conductance

Table 2.2 Critical Micelle Concentrations of Some Surface Active Agents

Surface active agent	cmc (mol dm^{-3})
(A) Anionic	
Sodium octyl-l-sulfate	1.30×10^{-1}
Sodium decyl-l-sulfate	3.32×10^{-2}
Sodium dodecyl-l-sulfate	8.39×10^{-3}
Sodium tetradecyl-l-sulfate	2.05×10^{-3}
(B) Cationic	
Octyl trimethylammonium bromide	1.30×10^{-1}
Decyl trimethylammonium bromide	6.46×10^{-2}
Dodecyl trimethylammonium bromide	1.56×10^{-2}
Hexadecyltrimethylammonium bromide	9.20×10^{-4}
(C) Nonionic	
Octyl hexaoxyethylene glycol monoether C_8E_6	9.80×10^{-3}
Decyl hexaoxyethylene glycol monoether $C_{10}E_6$	9.00×10^{-4}
Decyl nonaoxyethylene glycol monoether $C_{10}E_9$	1.30×10^{-3}
Dodecyl hexaoxyethylene glycol monoether $C_{12}E_6$	8.70×10^{-5}
Octylphenyl hexaoxyethylene glycol monoether C_8–φ–E_6	2.05×10^{-4}

of the surface active agent solution above the cmc, which is consistent with the reduction in mobility of the micelles as a result of counterion association. The presence of micelles also accounts for the rapid increase in light scattering or turbidity above the cmc

The presence of micelles was originally suggested by McBain [5] who suggested that below the cmc most of the surfactant molecules are unassociated, whereas in the isotropic solutions immediately above the cmc, micelles and surfactant ions (molecules) are thought to coexist, with the concentration of the latter changing very slightly as more surfactant is dissolved. However, the self-association of an amphiphile occurs in a stepwise manner with one monomer added to the aggregate at a time. For a long-chain amphiphile, the association is strongly cooperative up to a certain micelle size where counteracting factors became increasingly important [2]. Typically the micelles have a closely spherical shape in a rather wide concentration range above the cmc. Originally, it was suggested by Adam (6) and Hartley [7] that micelles are spherical in shape and have the following properties: (1) the association unit is spherical with a radius approximately equal to the length of the hydrocarbon chain; (2) the micelle contains about 50–100 monomeric units—aggregation number generally increases with increase in alkyl chain length; (3) with ionic surfactants, most counterions are bound to the micelle surface, thus significantly reducing the mobility from the value to be expected from a micelle with noncounterion bonding; (4) micellization occurs over a narrow concentration range as a result of the high association number of surfactant micelles; (5) the interior of the surfactant micelle has essentially the properties of a liquid hydrocarbon. This is confirmed by the high mobility of the alkyl chains and the ability of the micelles to solubilize many water insoluble organic molecules, e.g., dyes and agrochemicals.

Although, the spherical micelle model accounts for many of the physical properties of solutions of surfactants, a number of phenomena remain unexplained, without consideration of other shapes. For example, McBain [8] suggested the presence of two types of micelles—spherical and lamellar—in order to account for the drop in molar conductance of surfactant solutions. The lamellar micelles are neutral and hence account for the reduction in conductance. Later, Harkins et al. [9] used McBain's model of lamellar micelles to interpret his x-ray results in soap solutions. Moreover, many modern techniques such as light scattering and neutron scattering indicate that in many systems the micelles are not spherical. For example, Debye and Anacker [10] proposed a cylindrical micelle to explain that light scattering results on hexadecyltrimethylammonium bromide in water. Evidence for disc-shaped micelles have also been obtained under certain conditions. A schematic representation of the spherical, lamellar, and rod-shaped micelles, suggested by McBain, Hartly, and Debye, is given in Figure 2.2.

One of the characteristic features of solutions of surfactants is their solubility–temperature relationship, which is strikingly different for ionic and nonionic surfactants. With ionic surfactants, the solubility

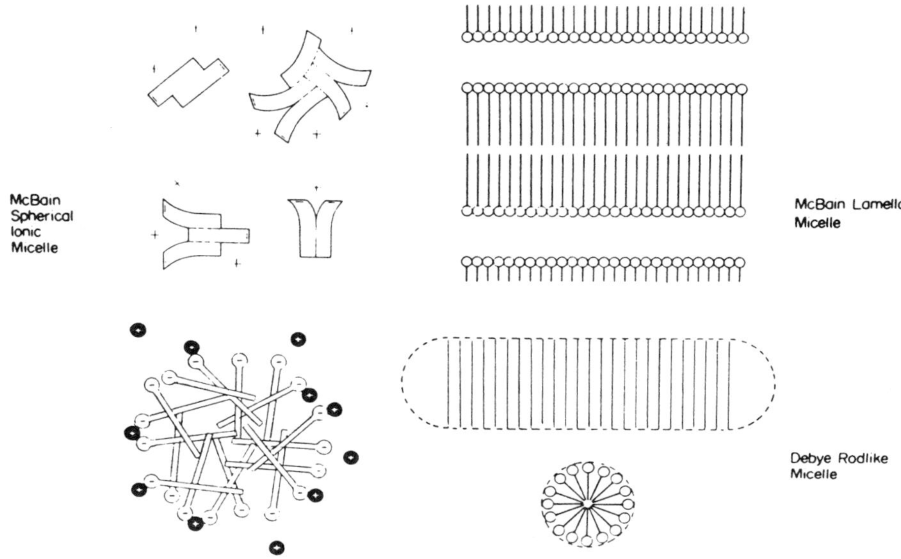

Figure 2.2 Various shapes of micelles, following McBain [5, 8], Hartley [7] and Debye [10].

first increases gradually with increase of temperature; then, above a certain temperature, there is a very sudden increase of solubility with further increase in temperature [11]. This is illustrated in Figure 2.3, which shows the results for sodium decyl sulfonate in water. The same figure also shows the variation of cmc with temperature [12]. It can be seen that the solubility of the surfactant increases rapidly above 22°C [12]. The cmc increases gradually with increase in temperature. At a particular temperature, the solubility becomes equal to the cmc, i.e., the solubility curve intersects the cmc. This temperature is referred to as the Krafft temperature of the surfactant, which for sodium decyl sulfate is 22°C. It should be mentioned that the Krafft temperature increases with increase of the alkyl chain length of the surfactant molecule.

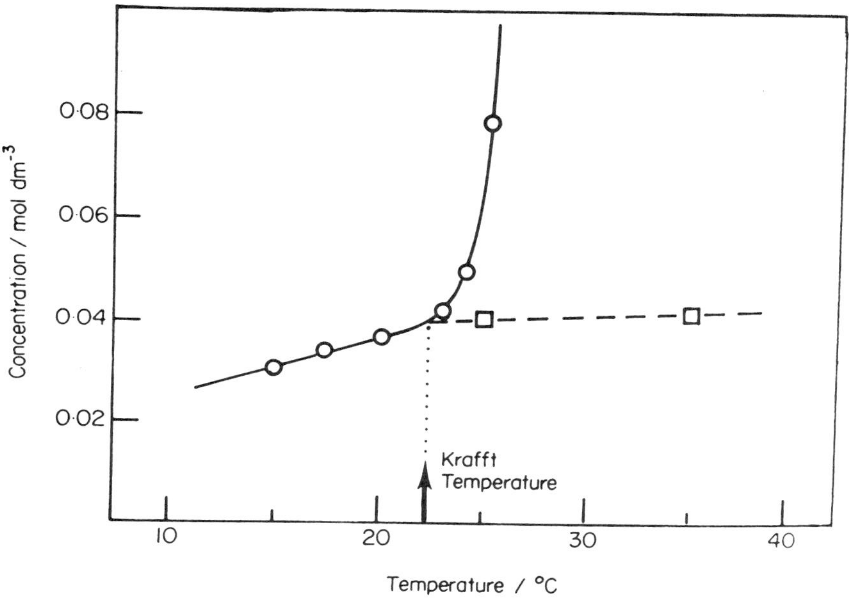

Figure 2.3 Solubility and cmc. Temperature for sodium decyl sulfonate in water. ○, solubility; □, cmc

The solubility–temperature relationship for nonionic surfactants is different from that for ionic surfactants. This is illustrated in Figure 2.4 which shows the phase diagram for the binary system dodecyl hexaoxyethylene glycol monoether–water [13]. This phase diagram shows the various phases that are formed when the surfactant concentration and temperature is changed. Let us first consider a dilute nonionic surfactant solution, say 1%, this solution is isotropic (denoted by I) at low temperatures, but on increasing the temperature, a critical point is reached above which the solution becomes turbid. This critical temperature is defined as the cloud point of the surfactant at this particular concentration. On further heating the solution above its cloud point, it separates into two liquid layers (defined by 2L), one

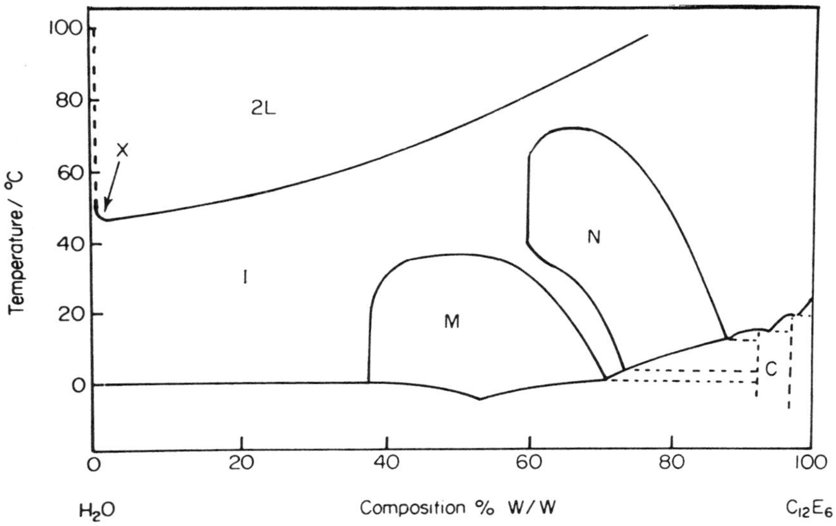

Figure 2.4 Phase diagram for dodecyl hexaoxyethylene glycol monoether–water mixture; I, Isotropic solution; 2L, two liquid phase; M, middle phase; N, neat phase; C, crystalline phase.

rich in water and one rich in surfactant. Thus, the line that separates the 2L from the isotropic solution (I) may be defined as the cloud point curve. It can be seen that the phase separation that first decreases with increase in surfactant concentration (in the dilute region) reaches a minimum (which may be defined as the lower consolute temperature) and then increases. The point x is characteristic of a nonionic surfactant–water mixture and hence may be defined as the cloud point of that particular solution. In most trade literature of surfactants, a cloud point is defined at a particular surfactant concentration (usually 1%). As is clear from Figure 2.4, the cloud point depends on the surfactant concentration, which needs to be specified to have any meaning. It is sometimes qualitatively stated that the solubility of nonionic surfactants decreases with increase of temperature using cloudiness or phase separation as the solubility limit. Strictly speaking, this is an incorrect

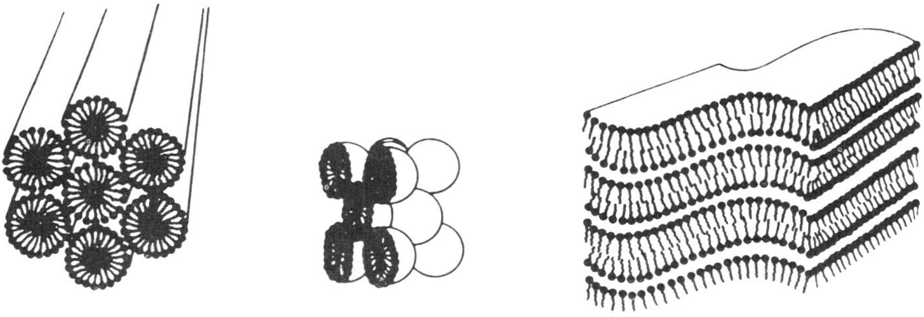

Figure 2.5 Schematic representation of structures found in concentrated surfac active agent solutions.

statement since that depends on which side of the minimum in the consolute boundary one is (Fig. 2.4).

The phase diagram of Figure 2.4 shows some characteristic regions at high surfactant concentrations, namely, the M and N region. The M phase is the region of the hexagonal or middle phase, consisting of cylindrical units that are hexagonally close-packed. In this region, the viscosity of the surfactant solution is extremely high and the system appears as a transparent gel. It shows characteristic textures under polarized light and hence the middle phase may be identified by optical microscopy by referring to published pictures (14). The N phase is the lamellar or neat phase, consisting of sheets of molecules in a bimolecular packng with head group exposed to the water layers in between them. This is less viscous than the M phase and it shows different textures under polarized light [14]. Several other liquid crystalline phases may be identified with other nonionic surfactant systems such as the cubic viscous isotropic phase. An illustration of the above mentioned three phases was given by Corkill and Goodman [15] (as shown in Fig. 2.5).

It should be mentioned that the concentration and temperature domains on which these phases are formed, the so-called lyotropic mesomorphic phases or lydropic liquids crystals, vary widely for different surfactants. Major changes also occur on addition of electrolytes or another organic phase as a long chain alcohol.

III. THERMODYNAMICS OF MICELLIZATION

As mentioned above, the process of micellization is one of the most important characteristics of surfactant solution and hence it is essential to understand its mechanism (the driving force for micelle formation). This requires analysis of the dynamics of the process (i.e., the kinetic aspects) as well as the equilibrium aspects whereby the laws of thermodynamics may be applied to obtain the free energy, enthalpy, and entropy of micellization. Below a brief description of both aspects will be given and this will be followed by a picture of the driving force for micelle formation.

A. Kinetic Aspects

Micellization is a dynamic pheonmenon in which n monomeric surfactant molecules associate to form a micelle S_n, i.e.,

$$n\,S \leftrightarrows S_n \tag{1}$$

Hartley [7] envisaged a dynamic equilibrium whereby surface active agent molecules are constantly leaving the micelles while other molecules from solution enter the micelles. The same applies to the counterions with ionic surfactants, which can exchange between the micelle surface and bulk solution.

Experimental investigation using fast kinetic methods such as stop flow, temperature and pressure jumps, and ultrasonic relaxation measurements have shown that there are two relaxation processes for micellar equilibrium [16–20] characterized by relaxation times τ_1 and τ_2. The first relaxation time, τ_1, is of the order of 10^{-7} s (10^{-8} to 10^{-3} s) and represents the lifetime of a surface active molecule in a micelle, i.e., it represents the association and dissociation rate for a single molecule entering and leaving the micelle, which may be represented by the equation,

$$S + S_{n-1} \leftrightarrows S_n \tag{2}$$

where K^+ and K^- represent the association and dissociation rate respectively for a single molecule entering or leaving the micelle.

The slower relaxation time τ_2 corresponds to a relatively slow process, namely, the micellization–dissolution process represented by Eq. (1). The value of τ_2 is of the order of milleseconds (10^{-3}–1 s) and hence can be conveniently measured by stopped flow methods. The fast relaxation time τ_1 can be measured using various techniques depending on its range. For example, τ_1 values in the range of 10^{-8}–10^{-7} s are accessible to ultrasonic absorption methods, whereas τ_1 in the range of 10^{-5}–10^{-3} s can be measured by pressure jump methods. The value of τ_1 depends on surfactant concentration, chain length, and temperature. τ_1 increases with increase of chain length of surfactants, i.e., the residence time increases with increase of chain length.

The above discussion emphasizes the dynamic nature of micelles and it is important to realize that these molecules are in continuous motion and that there is a constant interchange between micelles and solution. The dynamic nature also applies to the counterions, which exchange rapidly with lifetimes in the range 10^{-9}–10^{-8} s. Furthermore, the counterions appear to be laterally mobile and not to be associated with (single) specific groups on the micelle surfaces [2].

B. Equilibrium Aspects: Thermodynamic Treatment of Micellization

Two general approaches have been employed in tackling the problem of micelle formation. The first and most simple approach treats micelles as a single phase, and this is referred to as the phase separation model. In this model, micelle formation is considered as a phase separation phenomenon. The cmc is then the saturation concentration of the amphiphile in the monomeric state whereas the micelles constitute the separated pseudophase. Above the cmc, a phase equilibrium exists with a constant activity of the surfactant in the micellar phase. The Krafft point is viewed as the temperature at which solid hydrated surfactant, micelles and a solution saturated with undissociated surfactant molecules are in equilibrium at a given pressure.

In the second approach, micelles and single surfactant molecules or ions are considered to be in association–dissociation equilibrium. In its simplest form, a single equilibrium constant is used to treat the process represented by Eq. (1). The cmc is merely a concentration range above which any added surfactant appears in solution in a micellar form. Since the solubility of the associated surfactant is much greater than that of the monomeric surfactant, the solubility of the surfactant as a whole will not increase markedly with temperature until it reaches the cmc region. Thus, in the mass action approach, the Krafft point represents the temperature at which the surfactant solubility equals the cmc.

1. Phase Separation Model

Consider an anionic surfactant, in which n surfactant anions, S^-, and n counterions M^+ associate to form a micelle, i.e.,

$$n\,S^- + nM^+ \leftrightarrows S_n \tag{3}$$

The micelle is simply a charged aggregate of surfactant ions plus an equivalent number of counterions in the surrounding atmosphere, and is treated as a separate phase.

The chemical potential of the surfactant in the micellar state is assumed to be constant, at any given temperature, and this may be adopted as the standard chemical potential, μ_m^o, by analogy to a pure liquid or a pure solid. Considering the equilibrium between micelles and monomer, then

$$\mu_m^o = \mu_1^o + RT \ln a_1 \tag{4}$$

where μ_1 is the standard chemical potential of the surfactant monomer and a_1 is its activity which is equal to $f_1 x_1$, where f_1 is the activity coefficient and x_1 the mole fraction. Therefore, the standard free energy of micellization per mol of monomer, ΔG_m^o, is given by

$$\begin{aligned}\Delta G_m^o &= \mu_m^o - \mu_1^o \\ &= RT \ln a_1 \\ &\simeq RT \ln x_1 \end{aligned} \qquad (5)$$

where f_1 is taken as unity (a reasonable value in very dilute solution). The cmc may be identified with x_1 so that

$$\Delta G_m^o = RT \ln [\text{cmc}] \qquad (6)$$

In Eq. (6), the cmc is expressed as a mole fraction, which is equal to $C/(55.5 + C)$, where C is the concentration of surfactant in mol dm^{-3}, i.e.,

$$\Delta G_m^o = RT \ln C - RT \ln (55.5 + C) \qquad (7)$$

It should be stated that ΔG^o should be calculated using the cmc expressed as a mole fraction as indicated by Eq. (5). However, most cmc quoted in the literature are given in mol dm^{-3} and in many cases of ΔG^o values have been quoted when the cmc was simply expressed in mol dm^{-3}. Strictly speaking, this is incorrect, since ΔG^o should be based on x_1 rather than on C. The value of ΔG^o, when the cmc is expressed in mol dm^{-3}, is substantially different from the ΔG^o value when the cmc is expressed in mole fraction. For example, with dodecyl hexaoxyethylene glycol, the quoted cmc value is 8.7×10^{-5} mol dm^{-3} at 25°C. Therefore,

$$\begin{aligned}\Delta G^o &= RT \ln \frac{8.7 \times 10^{-5}}{55.5 + 8.7 \times 10^{-5}} \\ &= -33.1 \text{ kJ mol}^{-1}\end{aligned} \qquad (8)$$

when the mole fraction scale is used. On the other hand,

$$\Delta G^o = RT \ln 8.7 \times 10^{-5}$$
$$= -23.2 \text{ kJ mol}^{-1} \tag{9}$$

when the molarity scale is used.

The phase separation model has been questioned for two main reasons. First, according to this model a clear discontinuity in the physical property of a surfactant solution, such as surface tension, turbidity, etc., should be observed at the cmc. This is not always found experimentally and the cmc is not a sharp break point. Second, if two phases actually exist at the cmc, then equating the chemical potential of the surfactant molecule in the two phases would imply that the activity of the surfactant in the aqueous phase would be constant above the cmc. If this was the case, the surface tension of a surfactant solution should remain constant above the cmc. However, careful measurements have shown that the surface tension of a surfactant solution decreases slowly above the cmc, particularly when using purified surfactants.

2. Mass Action Model

This model assumes a dissociation–association equilibrium between surfactant monomers and micelles and an equilibrium constant can be calculated. For a nonionic surfactant, where charge effects are absent, this equilibrium is simply represented by Eq. (1) which assumes a single equilibrium. In this case, the equilibrium constant K_m is given by the equation,

$$K_m = \frac{[S_n]}{[S]^n} \tag{10}$$

The standard free energy per monomer is then given by

$$-\Delta G_m^o = \frac{\Delta G}{n}$$

$$= \frac{RT}{n} \ln K_m$$

$$= \frac{RT}{n} \ln [S_n] - RT \ln [S] \tag{11}$$

For many micellar systems, $n > 50$ and, therefore, the first term on the right hand side of Eq. (9) may be neglected, resulting in the following expression for ΔG^o_m,

$$\Delta G^o_m = RT \ln [S]$$

$$= RT \ln [\text{cmc}] \tag{12}$$

which is identical to the equation derived using the phase separation model.

The mass action model allows a simple extension to be made to the case of ionic surfactants, in which micelles attract a substantial proportion of counterions, into an attached layer. For a micelle made of n-surfactant ions, (where $n - p$ charges are associated with counterions, i.e., having a net charge of p units and degree of dissociation p/n), the following equilibrium may be established (for an anionic surfactant with Na^+ counterions):

$$n\, S^- + (n - p)\, Na^+ \leftrightarrows S_n^{p-} \tag{13}$$

$$K_m = \frac{[S_n^{p-}]}{[S^-]^n\, [Na^+]^{(n-p)}} \tag{14}$$

A convenient solution for relating ΔG_m to [cmc] was given by Phillips [21] who arrived at the following expression,

$$\Delta G^o_m = \left(2 - \frac{p}{n}\right) RT \ln [\text{cmc}] \tag{15}$$

For many ionic surfactants, the degree of dissociation (p/n) is about 0.2 so that

$$\Delta G^o_m = 1.8\, RT \ln [\text{cmc}] \qquad (16)$$

Comparison with Eq. (10) clearly shows that for similar ΔG_m, the [cmc] is about two orders of magnitude higher for ionic surfactants when compared with nonionic surfactant of the same alkyl chain length (see Table 2.2).

In the presence of excess added electrolyte, with mole fraction x, the free energy of micellization is given by the expression,

$$\Delta G^o_m = RT \ln [\text{cmc}] + \left(1 - \frac{p}{n}\right) \ln x \qquad (17)$$

Equation (15) shows that as x increases, the [cmc] decreases.

It is clear from Eq. (13) that as $p \rightarrow 0$, i.e., when most charges are associated with counterions,

$$\Delta G^o_m = 2\, RT \ln [\text{cmc}] \qquad (18)$$

whereas when $p \sim n$, i.e., the counterions are bound to micelles,

$$\Delta G^o_m = RT \ln [\text{cmc}] \qquad (19)$$

which is the same equation for nonionic surfactants.

Although the mass action approach could account for a number of experimental results such as small change in the properties around the cmc, it has not escaped criticism. For example, the assumption that surfactants exist in solution in only two forms, namely, single ions and micelles of uniform size, is debatable. Analysis of various experimental results has shown that micelles have a size distribution that is narrow and concentration dependent. Thus, the assumption of a single aggregation number is an oversimplification and in reality there is a micellar size distribution. This can be analyzed using the multiple equilibrium model which can be best formulated as a stepwise aggregation [2],

$$S_1 + S_1 \leftrightarrows S_2 \tag{20}$$

$$S_2 + S_1 \leftrightarrows S_3 \tag{21}$$

$$S_{n-1} + S_1 \leftrightarrows S_n \tag{22}$$

As noted in particular in the analysis of kinetic data [17, 22] there are aggregates over a wide range of aggregation numbers, from dimers and well beyond the most stable micelles. However, for surfactants with not too high cmc, the size distribution curve has a very deep minimum, the least stable aggregates being present in concentrations many orders of magnitude below those of the most abundant micelles. For surfactants with predominantly spherical micelles, the polydispersity is low and there is then a particularly preferred micellar size.

C. Enthalpy and Entropy of Micellization

The enthalpy of micellization can be calculated from the variation of cmc with temperature. This follows from

$$-\Delta H^o = RT^2 \frac{d \ln [\text{cmc}]}{dT} \tag{23}$$

The entropy of micellization can then be calculated from the relationship between ΔG^o and ΔH^o, i.e.,

$$\Delta G^o = \Delta H^o - T\Delta S^o \tag{24}$$

Therefore ΔH^o may be calculated from the surface tension–log C plots at various temperatures. Unfortunately, the errors in locating the cmc (which in many cases is not a sharp point) leads to a large error in the value of ΔH^o. A more accurate and direct method of obtaining ΔH^o is microcalorimetry. As an illustration, the thermodynamic para-

Table 2.3 Thermodynamic Quantities for Micellization of Octylhexaoxyethylene Glycol Monoether (kJ mol^{-1})

Temp (°C)	ΔG^o	ΔH^o (from cmc)	ΔH^o (from calorimitry)	
25	-21.3 ± 2.1	8.0 ± 4.2	20.1 ± 0.8	41.8 ± 1.0
40	-23.4 ± 2.1		14.6 ± 0.8	38.0 ± 1.0

meters, ΔG^o, ΔH^o, and $T\Delta S^o$ for octylhexaoxyethylene glycol monoether (C_8E_6) are given in Table 2.3. It can be seen from the table that ΔG^o is large and negative. However, ΔH^o is positive, indicating that the process is endothermic. In addition, $T\Delta S^o$ is large and positive which implies that in the micellization process there is a net increase in entropy. As we will see in the next section, this positive enthalpy and entropy points to a different driving force for micellization from that encountered in many aggregation processes.

The influence of alkyl chain length of the surfactant on the free energy, enthalpy and entropy of micellization was demonstrated by Rosen [23] who listed these parameters as a function of alkyl chain length for sulfoxide surfactants. The results are given in Table 2.4. It

Table 2.4 Change of Thermodynamic Parameters of Micellization of Alkyl Sulfoxide with Increasing Chain Length of the Alkyl Group (kJ mol^{-1})

Surfactant	ΔG	ΔH^o	$T\Delta S^o$
$C_6H_{13}S(CH_3)O$	−12.0	10.6	22.6
$C_7H_{15}S(CH_3)O$	−15.9	9.2	25.1
$C_8H_{17}S(CH_3)O$	−18.8	7.8	26.4
$C_9H_{19}S(CH_3)O$	−22.0	7.1	29.1
$C_{10}H_{21}S(CH_3)O$	−25.5	5.4	30.9
$C_{11}H_{23}S(CH_3)O$	−28.7	3.0	31.7

can be seen that the standard free energy of micellization becomes increasingly negative as the chain length increases. This is to be expected since the cmc decreases with increase of the alkyl chain length. However, ΔH^o becomes less positive and $T\Delta S$ becomes more positive with increase in chain length of the surfactant. Thus, the large negative free energy of micellization is made up of a small positive enthalpy (which decreases slightly with increase of the chain length of the surfactant) and a large positive entropy term $T\Delta S^o$, which becomes more positive as the chain is lengthened. As we will see in the next section, these results can be accounted for in terms of the hydrophobic effect, which will be described in some detail.

D. Driving Force for Micelle Formation

Until recently, the formation of micelles was regarded primarily as an interfacial energy process, analogous to the process of coalescence of oil droplets in an aqueous medium. If this was the case, micelle formation would be a highly exothermic process, as the interfacial free energy has a large enthalpy component. As mentioned above, experimental results have clearly shown that micelle formation involves only a small enthalpy change and is often endothermic. The negative free energy of micellization is the result of a large positive entropy. This led to the conclusion that micelle formation must be a predominantly entropy driven process. Two main sources of entropy may have been suggested. The first is related to the so-called hydrophobic effect. The latter effect was first established from a consideration of the free energy enthalpy and entropy of transfer of hydrocarbon from water to a liquid hydrocarbon. Some results are listed in Table 2.5. This table also lists the heat capacity change ΔC_p on transfer from water to a hydrocarbon, as well as $C_p^{o,\text{gas}}$, i.e., the heat capacity in the gas phase [2]. It can be seen from Table 2.5 that the principal contribution to the value of ΔG^o is the large positive value of ΔS^o, which increases with increase in the hydrocarbon chain length, whereas ΔH^o is positive, or small and negative. To account for this large positive entropy of transfer several authors [23, 24] suggest that the water molecules around a hydrocarbon chain are

Table 2.5 Thermodynamic Parameters for Transfer of Hydrocarbons from Water to Liquid Hydrocarbon at 25°C (kJ mol^{-1})

Hydrocarbon	ΔG^o	ΔH^o	ΔS^o	ΔC_p^o	$\Delta C_p^{o,gas}$
C_2H_6	−16.4	10.5	88.2	–	–
C_3H_8	−20.4	7.1	92.4	–	–
C_4H_{10}	−24.8	3.4	96.6	−273	−143
C_5H_{12}	−28.8	2.1	105.0	−403	−172
C_6H_{14}	−32.5	0	109.2	−441	−197
C_6H_6	−19.3	−2.1	58.8	−227	−134
$C_6H_5CH_3$	−22.7	−1.7	71.4	−265	−155
$C_6H_5C_2H_5$	−26.0	−2.0	79.8	−319	−185
$C_6H_5C_3H_8$	−29.0	−2.3	88.2	−395	–

ordered, forming "clusters" or "icebergs." On transfer of an alkane from water to a liquid hydrocarbon, these clusters are broken thus releasing water molecules that now have a higher entropy. This accounts for the large entropy of transfer of an alkane from water to a hydrocarbon medium. This effect is also reflected in the much higher heat capacity change on transfer, ΔC_p^o, when compared with the heat capacity in the gas phase, C_p^o.

The above effect is also operative on transfer of surfactant monomer to a micelle during the micellization process. The surfactant monomers will also contain "structured" water around their hydrocarbon chain. On transfer of such monomers to a micelle, these water molecules are released and have a higher entropy.

The second source of entropy increase on micellization may arise from the increase in flexibility of the hydrocarbon chains on their transfer from an aqueous to a hydrocarbon medium [25, 26]. The orientations and bendings of an organic chain are likely to be more restricted in an aqueous phase than to an organic phase.

It should be mentioned that with ionic and zwitterionic surfactants, an additional entropy contribution, associated with the ionic head groups, must be considered. Upon partial neutralization of the ionic charge by the counterions when aggregation occurs, water molecules are released. This will be associated with an entropy increase that should be added to the entropy increase resulting from the hydrophobic effect mentioned above. However, the relative contribution of the two effects is difficult to assess in a quantitative manner.

3

Adsorption of Surfactants and Polymers at the Air/Liquid, Liquid/Liquid, and Solid/Liquid Interfaces

As mentioned in the Introduction, surfactants play a major role in the formulation of agrochemicals. In the first place, they are used for stabilization of disperse systems such as emulsions, suspensions, and microemulsions. Second, surfactants are added in emulsifiable concentrates for their spontaneous dispersion on dilution. This is also the case with wettable powders in which dispersion of the solid particle requires wetting and spontaneous breaking of any aggregates or agglommerates, their subsequent stabilization, and also for application to enhance wetting. Apart from their roles as stabilizers, surfactants also play a major role in the transfer of the chemical.

In all the above-mentioned phenomenon, the surfactant needs to accummulate at the interface, a process that is generally described as adsorption. The simplest interface is that of the air/liquid and in this case the surfactant will adsorb with the hydrophilic group pointing toward the polar liquid (water) leaving the hydrocarbon chain pointing toward the air. This process results in lowering of the surface tension γ. Typically, surfactants show a gradual reduction in γ, until the cmc is reached above which the surface tension remains virtually constant. Hydrocarbon surfactants of the ionic, nonionic, or Zwitterionic type lower the surface tension to limiting values reaching 30–40 mN m^{-1} depending on the nature of the surfactant. Lower values may be achieved using fluorocarbon surfactants, typically of the order of 20 mN m^{-1}. As we will see in the last chapter of this book, this surface tension lowering is essential for spray application. The surfactants enable the disruption and atomization of the liquid by lowering γ and inducing a surface tension gradient. It is, therefore, essential to understand the adsorption and conformation of surfactants at the air/liquid interface.

With emulsifiable concentrates, emulsions, and microemulsions, the surfactant adsorbs at the oil/water interface, with the hydrophilic head group immersed in the aqueous phase, leaving the hydrocarbon chain in the oil phase. Again, the mechanism of stabilization of emulsions and microemulsions depends on the adsorption and orientation of the surfactant molecules at the liquid/liquid interface. As we will see in subsequent chapters, macromolecular surfactants (polymers) are nowadays being used for stabilization of emulsions and hence it is essential to understand their adsorption at the interface. The first part of this chapter will deal with the adsorption of surfactants at the air/liquid and liquid/liquid interfaces, since the same equations can be applied for treating adsorption of surfactants at both these interfaces. The second part of this chapter will deal with the adsorption of surfactants and polymers at the solid/liquid interface. This is key to understanding how surfactants and polymers function as stabilizers and flocculants. Since surfactant and polymer adsorption processes are significantly different, the two subjects will be treated separately. Suffice to say at this stage that surfactant adsorption is relatively more simple than polymer adsorption. This stems from the fact that surfactants consist of a small number of units and they are mostly reversibly adsorbed, allowing one to apply some thermodynamic treatments. In this case, it is possible to describe the adsorption in terms of various interaction parameters such as chain surface, chain solvent, and surface solvent. Moreover, the configuration of the surfactant molecule can be simply described in terms of these possible interactions. In contrast, the process of polymer adsorption is fairly complicated. In addition to the usual adsorption considerations described above, one of the principle problems to be resolved is the configuration of the polymer molecule on the surface. This can be obtained in various possible ways depending on the number of segments and chain flexibility.

I. ADSORPTION OF SURFACTANTS AT THE AIR/LIQUID AND LIQUID/LIQUID INTERFACE

Before describing surfactant adsorption at the air/liquid (A/L) and liquid/liquid (L/L) interface it is essential to define the interface. The surface of a liquid is the boundary between two bulk phases, namely, liquid and air (or the liquid vapor). Similarly an interface between two immiscible liquids (oil and water) may be defined providing a dividing line is introduced since the interfacial region is not a layer that is one molecule thick, but usually has a thickness δ with properties that are different from the two bulk phases α and β [27]. However, Gibbs [28] introduced the concept of a mathematical dividing plane Z_σ in the interfacial region (Fig. 3.1). In this model the two

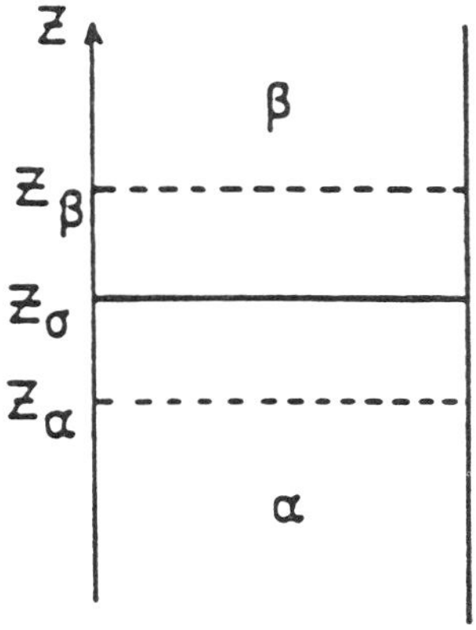

Figure 3.1 Gibbs convention for an interface.

bulk phases α and β are assumed to have uniform thermodynamic properties up to Z_σ. This picture applies for both the A/L and L/L interface (with A/L interfaces, one of the phase is air saturated with the vapor of the liquid).

Using the Gibbs model, it is possible to obtain a definition of the surface or interfacial tension γ, starting from the Gibbs–Deuhem equation [28], i.e.,

$$dG^\sigma = -S^\sigma \, dT + A \, d\gamma + \sum n_i \, d\mu_i \tag{25}$$

where G^σ is the surface free energy, S^σ is the entropy, A is the area of the interface, n_i is the number of moles of component i with chemical potential μ_i at the interface. At constant temperature and composition of the interface (i.e., in absence of any adsorption),

$$\gamma = \left(\frac{\partial G^\sigma}{\partial A} \right)_{T,n_i} \tag{26}$$

It is obvious from Eq. (26) that for a stable interface γ should be positive. In other words, the free energy should increase if the area of the interface increases, otherwise the interface will become convoluted, increasing the interfacial area until the liquid evaporates (for A/L case) or the two "immiscible" phases dissolved in each other (for the L/L case).

It is clear from Eq. (26), that surface or interfacial tension, i.e., the force per unit length tangentialy to the surface measured in units of mN m^{-1}, is dimensionally equivalent to an energy per unit area measured in mJ m^{-2}. For this reason, it has been stated that the excess surface free energy is identical to the surface tension, but this is true only for a single component system, i.e., a pure liquid (where the total adsorption is zero).

There are generally two approaches for treating surfactant adsorption at the A/L and L/L interface. The first approach, adopted by Gibbs, treats adsorption as an equilibrium phenomenon whereby the second law of thermodynamics may be applied using surface quantities. The second approach, referred to as the equation of state approach, treats the surfactant film as a two-dimensional layer with a surface pressure π that may be related to the surface excess Γ (amount of surfactant adsorbed per unit area). These two approaches are summarized below.

A. The Gibbs Adsorption Isotherm

Gibbs [28] derived a thermodynamic relationship between the surface or interfacial tension γ and the surface excess Γ (adsorption per unit area). The starting point of this equation is the Gibbs–Deuhem equation given above [Eq. (25)]. At constant temperature, but in the presence of adsorption, Eq. (25) reduces to

$$d\gamma = -\sum \frac{n_i^\sigma}{A}\, d\mu_i$$

$$= -\sum \Gamma_i \, d\mu_i \tag{27}$$

where

$$\Gamma_i = \frac{n_i^\sigma}{A}$$

is the number of moles of component i and adsorbed per unit area.

Equation (27) is the general form for the Gibbs adsorption isotherm. The simplest case of this isotherm is a system of two components in which the solute (2) is the surface active component, i.e., it is adsorbed at the surface of the solvent (1). For such a case, Eq. (27) may be written as,

$$-d\gamma = \Gamma_1^\sigma \, d\mu_1 + \Gamma_2^\sigma \, d\mu_2 \tag{28}$$

and if the Gibbs dividing surface is used,

$$\Gamma_1 = 0$$

and

$$-d\gamma = \Gamma^{\sigma}_{1,2}\, d\mu_2 \tag{29}$$

where $\Gamma^{\sigma}_{2,1}$ is the relative adsorption of (2) with repsect to (1). Since,

$$\mu_2 = \mu_2^o + RT \ln a_2^L$$

or

$$d\mu_2 = RTd \ln a_2^L \tag{30}$$

then,

$$-d\gamma = \Gamma^{\sigma}_{2,1}\, RT\, d \ln a_2^L \tag{31}$$

or

$$\Gamma^{\sigma}_{2,1} = -\frac{1}{RT}\left(\frac{d\gamma}{d \ln a_2^L}\right) \tag{32}$$

where a_2^L is the activity of the surfactant in bulk solution that is equal to $C_2 f_2$ or $x_2 f_2$, where C_2 is the concentration of the surfactant in mol dm^{-3} and x_2 is its mole fraction.

Equation (31) allows one to obtain the surface excess (abbreviated as Γ_2) from the variation of surface or interfacial tension with surfactant concentration. Note that $a_2 \sim C_2$ since in dilute solutions $f_2 \sim 1$. This approximation is valid since most surfactants have low cmc (usually less than 10^{-3} mol dm^{-3}) but adsorption is complete at or just below the cmc.

The surface excess Γ_2 can be calculated from the linear portion of the $\gamma - \log C_2$ curves before the cmc. Such $\gamma - \log C$ curves are illustrated in Figure 3.2 for the air/water and oil/water interfaces; $[C_{SAA}]$ denotes the concentration of surface active agent in bulk solution. It can be seen that for the air/water interface γ decreases from

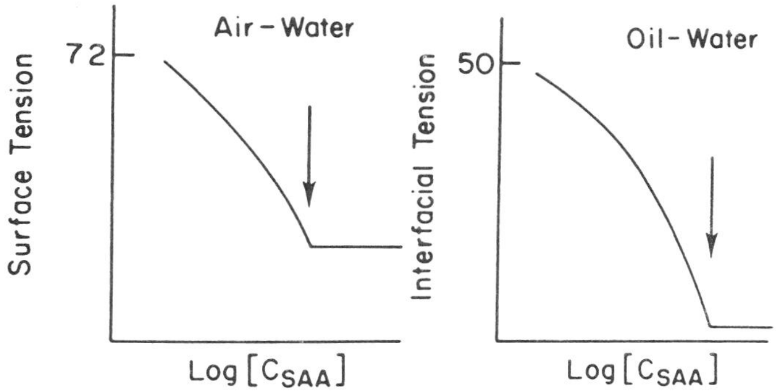

Figure 3.2 Variation of surface and interfacial tension with log [C_{SAA}] at the air/water and oil/water interface.

the value for water (72 mN m^{-1} at 20°C) reaching about 25–30 mN m^{-1} near the cmc. This is clearly schematic since the actual values depend on the surfactant nature. For the oil/water case, γ decreases from a value of about 50 mN m^{-1} (for a pure hydrocarbon-water interface) to about 1–5 mN m^{-1} near the cmc (again depending on the nature of the surfactant).

As mentioned above, Γ_2 can be calculated from the slope of the linear position of the curves shown in Figure 3.2 just before the cmc is reached. From Γ_2, the area per surfactant ion or molecule can be calculated since,

$$\text{Area/molecule} = \frac{1}{\Gamma_2 N_{av}} \tag{33}$$

where N_{av} is Avogadro's constant. Determining the area per surfactant molecule is very useful since it gives information on surfactant orientation at the interface. For example, for ionic surfactants such as sodium dodecyl sulfate, the area per surfactant is determined by the area occupied by the alkyl chain and head group if these molecules lie flat at the interface, whereas for vertical orientation, the area per surfactant ion is determined by that occupied by the charged head group, which at low electrolyte concentration will be in the region of 0.40 nm^2. Such area is larger than the geometrical area occupied by a sulfate group, as a result of the lateral repulsion between the head group. On addition of electrolytes, this lateral repulsion is reduced and the area/surfactant ion for vertical orientation will be lower than 0.4 nm^2 (in some cases reaching 0.2 nm^2). On the other hand, if the molecules lie flat at the interface the area per surfactant ion will be considerably higher than 0.4 nm^2.

Another important point can be made from the $\gamma - \log C$ curves. At concentration just before the break point, one has the condition

$$\left(\frac{\partial \gamma}{\partial \ln a_2}\right)_{p,T} = \text{constant} \tag{34}$$

which indicates that saturation adsorption has been reached. Just above the break point,

$$\left(\frac{\partial \gamma}{\partial \ln a_2}\right)_{p,T} = 0 \tag{35}$$

indicating the constancy of γ with $\log C$ above the cmc integration of Eq. (34) gives,

$$\gamma = \text{constant} \times \ln a_2 \tag{36}$$

Since γ is constant in this region, then a_2 must remain constant. This means that addition of surfactant molecules above the cmc must result in association to form units (micellar) with low activity.

As mentioned in Chapter 2, the hydrophilic head group may be un-ionized, e.g., alcohols or polyethylene oxide alkane or alkyl phenol compounds, weakly ionized such as carboxylic acids or strongly ionized such as sulfates, sulfonates, and quaternary ammonium salts. The adsorption of these different surfactants at the air/water and oil/ water interface depends on the nature of the head group. With nonionic surfactants, repulsion between the head groups is small and these surfactants are usually strongly adsorbed at the surface of water from very dilute solutions. As mentioned in Chapter 2, nonionic surfactants have much lower cmc values than with ionic surfactants with the same alkyl chain length. Typically, the cmc is in the region of 10^{-5}–10^{-4} mol dm^{-3}. Such nonionic surfactants form closely packed adsorbed layers at concentrations lower than their cmc values. The activity coefficient of such surfactants is close to unity and is only slightly affected by addition of moderate amounts of electrolytes (or change in the pH of the solution). Thus, nonionic surfactant adsorption is the simplest case since the solutions can be represented by a two component system and the adsorption can be accurately calculated using Eq. (31).

With ionic surfactants, on the other hand, the adsorption process is relatively more complicated since one has to consider the repulsion between the head groups and the effect of presence of any indifferent electrolyte. Moreover, the Gibbs adsorption equation has to be solved taking into account the surfactant ions, the counterion and any indifferent electrolyte ions present. For a strong surfactant electrolyte such as an Na^+R^-

$$\Gamma_2 = \frac{1}{2\,RT}\,\frac{\partial\gamma}{\partial \ln a\pm} \tag{37}$$

The factor of 2 in Eq. (37) arises because both surfactant ion and counter ion must be adsorbed to maintain neutrally, and $d\gamma/d \ln a\pm$ is twice as large as for an un-ionized surfactant.

If a nonadsorbed electrolyte, such as NaCl, is present in large excess, then any increase in concentration of Na^+R^- produces a negligible increase in Na^+ ion concentration and therefore $d\mu_{Na}$ becomes negligible. Moreover, $d\mu_{Cl}$ is also negligible, so that the Gibbs adsorption equation reduces to

$$\Gamma_2 = -\frac{1}{RT}\left(\frac{\partial\gamma}{\partial \ln C_{NaR}}\right) \tag{38}$$

i.e., it becomes identical to that for a nonionic surfactant.

The previous discussion, clearly illustrates that for calculation of Γ_2 from the $\gamma - \log C$ curve one has to consider the nature of the surfactant and the composition of the medium. For nonionic surfactants the Gibbs adsorption Eq. (31) can be directly used. For ionic surfactant, in the absence of electrolytes the right hand side of Eq. (31) should be divided by 2 to account for surfactant dissociation. This factor disappears in the presence of a high concentration of an indifferent electrolyte.

B. Equation of State Approach

In this approach, one relates the surface pressure π with the surface excess Γ_2. The surface pressure is defined by the equation,

$$\pi = \gamma_o - \gamma \tag{39}$$

where γ_o is the surface or interfacial tension before adsorption and γ that after adsorption.

For an ideal surface film, behaving as a two-dimensional gas the surface pressure π is related to the surface excess Γ_2 by the equation

$$\pi A = n_2 RT \tag{40}$$

or

$$\begin{aligned} \pi &= \frac{n_2}{A} RT \\ &= \Gamma_2 RT \end{aligned} \tag{41}$$

Differentiating Eq. (39) at constant temperature,

$$d\pi = RT\, d\Gamma_2 \tag{42}$$

Using the Gibbs equation,

$$d\pi = -d\gamma$$

$$= \Gamma_2 RT \, d \ln a_2$$
$$\simeq \Gamma_2 RT \, d \ln C_2 \quad (43)$$

Combining Eqs. (40) and (41)

$$d \ln \Gamma_2 = d \ln C_2 \quad (44)$$

or

$$\Gamma_2 = K \, C_2^{\alpha} \quad (45)$$

Equation (42) is referred to as the Henry's law isotherm which predicts a linear relationship between Γ_2 and C_2.

It is clear that Eqs. (39) and (42) are based on an idealized model in which the lateral interaction between the molecules has not been considered. Moreover, in this model the molecules are considered to be dimensionless. This model can only be applied at very low surface coverages where the surfactant molecules are so far apart that lateral interaction may be neglected. Moreover, under these conditions the total area occupied by the surfactant molecules is relatively small compared to the total interfacial area.

At significant surface coverages, the above equations have to be modified to take into account both lateral interaction between the molecules and the area occupied by them. Lateral interaction may reduce π if there is attraction between the chains (e.g., with most nonionic surfactant) or it may increase π as a result of repulsion between the head groups in the case of ionic surfactants.

Various equation of state have been proposed, taking into account the above two effects, in order to fit the $\pi - A$ data. The two-dimensional van der Waals equation of state is probably the most convenient for fitting these adsorption isotherms, i.e.,

$$\left(\pi + \frac{(n_2)^2 \, \alpha}{1 - \theta} \right) (A - n_2 A_2^{o}) = n_2 RT \quad (46)$$

where A_2^o is the excluded area or coarea of type 2 molecule in the interface and α is a parameter that allows for lateral interaction.

Equation (43) leads to the following theoretical adsorption isotherm, using the Gibbs equation:

$$C_2^\alpha = K_1 \left(\frac{\theta}{1-\theta}\right) \exp\left(\frac{\theta}{1-\theta} - \frac{2\alpha\theta}{a_2^o RT}\right) \tag{47}$$

where θ is the surface coverage,

$$\theta = \frac{\Gamma_2}{\Gamma_{2,\max}}$$

K_1 is constant that is related to the free energy of adsorption of surfactant molecules at the interface

$$-K_1 \quad \exp\left(-\Delta\frac{G_{\text{ads}}}{kT}\right)$$

and a_2^o is the area/molecule.

For a charged surfactant layer, Eq. (44) has to be modified to take into account the electrical contribution from the ionic head groups, i.e.,

$$C_2^\alpha = K_1 \left(\frac{\theta}{1-\theta}\right) \exp\left(\frac{\theta}{1-\theta}\right) \exp\left(\frac{e\Psi_o}{kT}\right) \tag{48}$$

where Ψ_o is the surface potential. Equation (45) shows how the electrical potential energy (Ψ_o/kt) of adsorbed surfactant ions affects the surface excess. Assuming that the bulk concentration remains constant, then Ψ_o increases as θ increases. This means that

$$\left[\frac{\theta}{1-\theta} \exp\left(\frac{\theta}{1-\theta}\right)\right]$$

increases less rapidly with C_2, i.e., adsorption is inhibited as a result of ionization.

II. ADSORPTION OF SURFACTANTS AND POLYMERS AT THE SOLID/LIQUID INTERFACE

As mentioned in the introduction, the adsorption of surfactant and polymers at the solid/liquid interface is vastly different and hence the two subjects will be treated separately.

A. Surfactant Adsorption at the Solid/Liquid Interface

In principle, surfactant adsorption may be described in terms of the surfactant/surface, surfactant/solvent, and surface/solvent interaction parameters. However, such interactions are not fully understood and they may involve ill-defined forces, e.g., hydrophobic bonding, solvation, chemisorption, etc. Moreover, the adsorption of ionic surfactants will involve electrostatic forces since solid surfaces acquire a charge in aqueous solution. For that reason, the adsorption of ionic and nonionic surfactants will be dealt with in separate sections.

1. Adsorption of Ionic Surfactants

Most pesticides are hydrophobic in nature and therefore the adsorption of ionic surfactants on their surfaces will be mainly governed by hydrophobic bonding and electrostatic interaction will play a relatively smaller role. However, if the surfactant head group is of the same charge sign as the particle surface, then electrostatic interaction opposes adsorption (due to repulsion), whereas if the head group has an opposite charge to the surface, electrostatic attraction enhances adsorption. Since the adsorption depends on the magnitude of hydrophobic bonding free energy, the amount of surfactant adsorbed increases directly with increase in the chain length of the surfactant in accordance with Traube's rule.

The adsorption of ionic surfactants on hydrophobic surfaces may be described using different equations. For example, the adsorption of ionic surfactants on hydrophobic surfaces may be represented by the Stern–Langmuir isotherm,

$$\frac{\theta}{1-\theta} = \frac{C}{55.51} \exp\left(\frac{-\Delta G^o_{ads}}{kT}\right) \tag{49}$$

where θ is the fraction surface coverage that is given by Γ/N_s, where Γ is the number of surfactant moles adsorbed per unit area and N_s is the total number of adsorption sites (in moles) per unit area for monolayer saturation adsorption, C is the bulk solution concentration, ΔG^o_{ads} is the adsorption free energy, k the Boltzmann constant, and T the absolute temperature.

An alternative expression that can describe surfactant adsorption is the Frumkin-Fowler-Guggenheim equation,

$$\frac{\theta}{1-\theta} \exp\,(A\theta) = \frac{C}{55.51} \exp\left(\frac{-\Delta G^o_{ads}}{kT}\right) \tag{50}$$

where A is a constant that is introduced to account for lateral interaction between the surfactant ions. At low coverage, A would reflect the repulsive electrostatic interaction, whereas at high coverage A reflects attractive chain–chain interaction.

Various contributions to the adsorption free energy may be envisaged. To a first approximation, these contributions may be considered to be additive. In the first instance, ΔG_{ads} may be taken to consist of two main contributions, i.e.,

$$\Delta G_{ads} = \Delta G_{elec} + \Delta G_{spec} \tag{51}$$

where ΔG_{elec} accounts for any electrical interactions and ΔG_{spec} is a specific adsorption term that contains all contributions to the adsorption free energy that are dependent on the "specific" (nonelectrical) nature of the system [29]. Several authors subdivided ΔG_{spec} into supposedly separate independent interactions [30–32], e.g.,

$$\Delta G_{spec} = \Delta G_{cc} + \Delta G_{cs} + \Delta G_{hs} + \cdots \tag{52}$$

where ΔG_{cc} is a term that accounts for the cohesive chain–chain interaction between the hydrophobic moieties of the adsorbed ions, ΔG_{cs} is the term for chain/substrate interaction whereas ΔG_{hs} is a term for the head group/substrate interaction. Several other contributions to ΔG_{spec} may be envisaged, e.g., ion–dipole, ion–induced dipole, or dipole–induced dipole interactions.

Only few results of adsorption isotherms of ionic surfactants on agrochemicals have been published in the literature. For example, Tadros [33] studied the adsorption of a cationic surfactant namely cetyl trimethylammonium bromide, an anionic surfactant, namely, sodium dodecyl benezenesulfonate (NaDBS) on a fungicide powder (ethirimol). Figure 3.3 shows the adsorption isotherm of CTABr on ethirimol. A two step isotherm with saturation adsorptions of 1.8–2.0 $\times 10^{-6}$ mol g^{-1} and 3.5 $\times 10^{-6}$ mol g^{-1} is clearly observed. Using the surface area of the powder obtained from krypton adsorption (0.29 m^2 g^{-1}), the area per CTA^+ ion was calculated to be 0.27 nm^2 and 0.14 nm^2 at the first and second plateau, respectively [33]. The stepwise isotherm could be due to bilayer adsorption of CTA^+ ion, the first layer with the charge pointing to the surface (which is negatively charged) and the second layer forming by hydrophobic bonding between the hydrocarbon chains of the surfactant ions. However, electrophoresis results showed that the isoelectric point occurs at much lower CTABr concentration (1 $\times 10^{-5}$ mol/100 ml). Thus adsorption probably occurs through a different mechanism to that described above. Initially, the ethirimol surface has a negative surface charge (as indicated by the negative ζ potential of −22 mV before addition of CTABr) which could be due to the presence of some specifically adsorbed anions. In the initial adsorption stage, these charges are neutralized with CTA^+ ions, with the positive charge pointing toward the surface, leaving uncovered hydrophobic regions in between. The subsequent CTA^+ ions will now adsorb on the hydrophobic sites, with the positive charge pointing toward the solution. The adsorption continues until a vertically oriented monolayer of

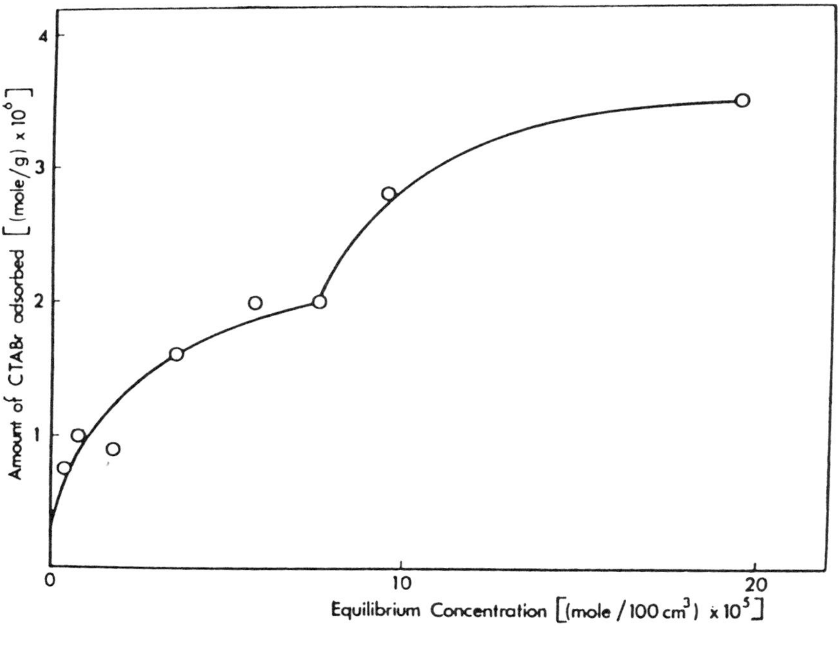

Figure 3.3 Adsorption isotherm of CTABr on ethirimol.

CTA^+ ions is reached at the first plateau. However, the area/CTA^+ ion of this plateau (0.27 nm^2) is lower than that expected from models of a close packed trimethylammonium ion (0.35 nm^2). Robb and Alexander [34] found on polyacrylonitrile latices an area of 0.35 nm^2/CTA^+ ion, whereas Conner and Ottewill [35] found an area of 0.47 nm^2 on polystyrene latex in 10^{-3} mol dm^{-3} KCl and 0.35 nm^2 in 5×10^{-2} mol dm^{-3} KBr. In view of the possible error in the gas adsorption surface area, the value of 0.27 nm^2 seems a reasonable estimate for the above-mentioned vertical orientation. As the surface of ethirimol becomes fully saturated with CTA^+ ions, there is still the possibility of association of CTA^+ ions on the surface that could result in a rapid increase in adsorption. This picture of association, which was originally suggested by Gaudin and Fuerstenau [36] could explain

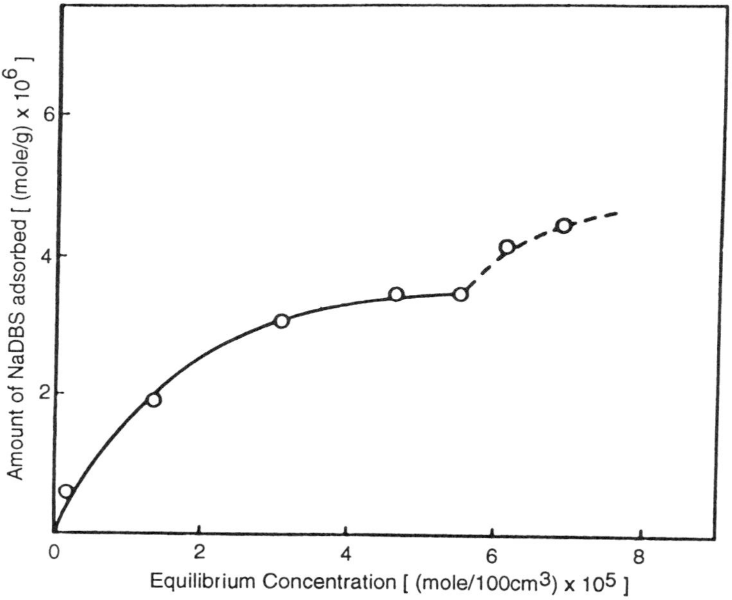

Figure 3.4 Adsorption isotherm of NaDBS on ethirimol.

the second plateau. These authors explained the rapid increase in terms of "hemimicellization." In other words, at a critical concentration (to be denoted cmc for hemimicelle formation), the hydrophobic moieties of the adsorbed surfactant are "squeezed out" from the aqueous solution by forming two-dimensional aggregates on the adsorbent surface. This is analogous to the process of micellization in bulk solution.

Figure 3.4 shows the adsorption isotherm of NaDBS on ethirimol. The isotherm is more or less of a Langmuir type with a saturation adsorption of 3.5×10^{-6} mol NaDBS/g ethirimol, which is similar to that obtained with CTABr at the second plateau. However, this saturation adsorption is reached at a much lower equilibrium adsorption of NaDBS and there is a tendency of further increase in adsorption beyond this value as indicated by the dotted line on the isotherm. Using the gas adsorption surface area, the area per DBS ion at the initial plateau is ~0.14 nm^2 which is smaller than that to be expected for a vertically oriented close packed monolayer of dodecyl benzenesulfonate ions. Again, this lower value could be accounted for in terms of the possible error in the gas adsorption surface area. The latter has been determined using krypton adsorption due to the small surface area and an error of more than 30% in the measured area is possible.

It is clear that the adsorption of an anionic surfactant such as NaDBS on the negative ethirimol surface should have occurred by hydrophobic bonding, i.e., the hydrophobic part of the molecule pointing toward the hydrophobic surface. Association or hemimicelle formation could account for the increased adsorption beyond the amount required to form a close packed monolayer.

2. Adsorption of Nonionic Surfactants

Several types of nonionic surfactants exist depending on the nature of the polar (hydrophilic) group. The most common type is that based on a polyoxyethylene glycol group, i.e., $(CH_2CH_2O)_nOH$ (where n can vary from as little as 2 units to as high as 100 or more units) linked either to an alkyl (C_xH_{2x+1}) or an alkylphenyl (C_xH_{2x+1}–C_6H_4–) group. These surfactants may be abbreviated as C_xE_n or $C_x\varphi E_n$ (where C refers to the number of C atoms in the alkyl chain, φ denotes C_6H_4 and E denotes ethylene oxide). These ethoxylated surfactants are characterized by a relatively large head group compared to the alkyl chain (when $n > 4$). However, there are nonionic surfactants with small head group such as amine oxides (-N $\rightarrow$ O) head group, phosphine oxide (-P $\rightarrow$ O) or sulfinylalkanol (SO-$(CH_2)_nOH$) (37). Most adsorption isotherms in the literature are based on the ethoxylated type surfactants.

The adsorption isotherms of nonionic surfactants are in many cases Langmuirian, like those of most other highly surface active solutes adsorbing from dilute solutions, and adsorption is generally reversible. However, several other adsorption types are produced [37] and these are illustrated in Figure 3.5. The steps in the isotherm may be explained in terms of the various adsorbate–adsorbate, adsorbate–adsorbant, and adsorbate–solvent interactions. These orientations are schematically illustrated in Figure 3.6. In the first stage of adsorption (denoted

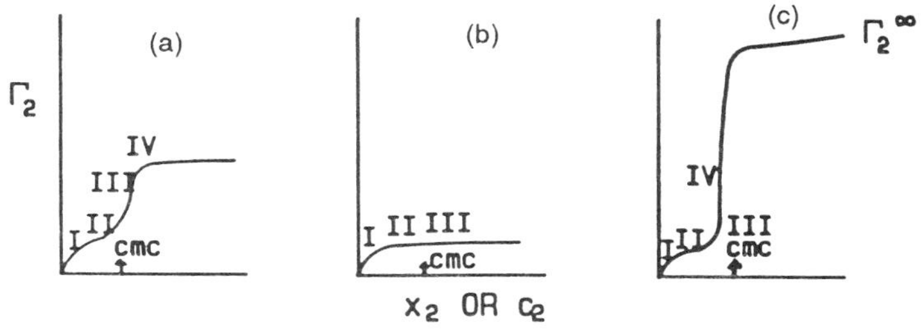

Figure 3.5 Adsorption isotherms, corresponding to the three adsorption sequences shown in Figure 3.6 (I–IV), indicating the different orientation; the cmc is indicated by an arrow.

by I in Figs. 3.5 and 3.6), surfactant–surfactant interaction is negligible (low coverage) and adsorption occurs mainly by van der Waals interaction. On a hydrophobic surface, the interaction is dominated by the hydrophobic portion of the surfactant molecule. This is mostly the case with agrochemicals with hydrophobic surfaces. However, if the chemical is hydrophilic in nature, the interaction will be dominated by the ethylene oxide (EO) chain. The approach to monolayer saturation with the molecules lying flat is accompanied by a gradual decrease in the slope of the adsorption isotherm (region II in Figure 3.5). An increase in the size of the surfactant molecule, e.g., increasing the length of the alkyl or EO chain, will decrease adsorption (when expressed in moles per unit area). On the other hand, increasing temperature will increase adsorption as a result of desolvation of the EO chains, thus reducing their size. Moreover, increasing temperature reduces the solubility of the nonionic surfactant and this enhances adsorption.

The subsequent stages of adsorption (regions III and IV) are determined by surfactant–surfactant interaction, although surfactant–surface interaction initially determines adsorption beyond stage II. This interaction depends on the nature of the surface and the hydrophilic–lipophilic balance (HLB) of the surfactant molecule. For a hydrophobic surface, adsorption occurs via the alkyl group of the surfactant. For a given EO chain, the adsorption will increase with an increase in the alkyl chain length. On the other hand, for a given lateral forces due

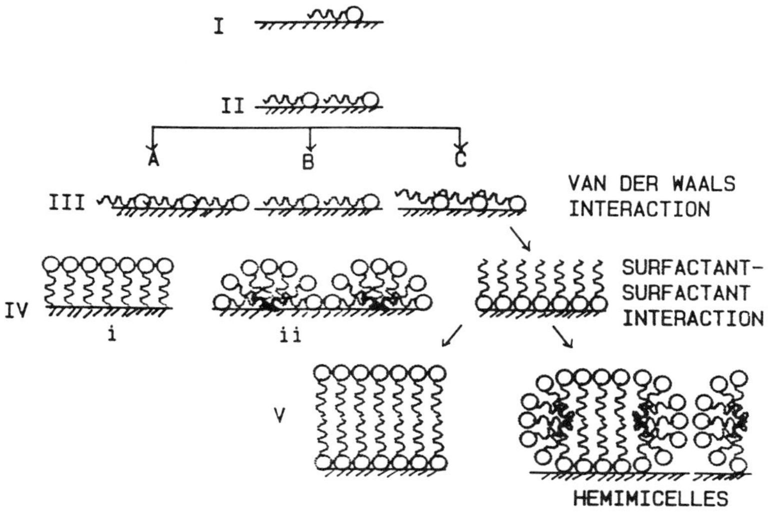

Figure 3.6 Model for the adsorption of nonionic surfactants showing orientation of surfactant molecules at the surface. I–V are the successive stages of adsorption, and sequence A–C corresponds to situations where there are relatively weak, intermediate, and strong interactions between the adsorbant and the hydrophilic moiety of the surfactant.

to alkyl chain interactions in the adsorbed layer alkyl chain length, adsorption increases with decreasing EO chain length. As the surfactant concentration approaches the cmc, there is a tendency for aggregation of the alkyl groups. This will cause vertical orientation of the surfactant molecules (stage IV). This will compress the head group and for an EO chain this will result in a less coiled more extended conformation. The larger the surfactant alkyl chain the greater will be the cohesive forces and hence the smaller the cross sectional area. This may explain why saturation adsorption increases with increasing alkyl chain length.

The interaction occurring in the adsorption layer during the fourth and subsequent stages of adsorption are similar to those that occur in bulk solution. In this case aggregate units, as shown in Figure 3.6 V (hemimicelles or micelles), may be formed. This picture was supported by Klimenker et al. [38] who found close agreement between saturation adsorption and adsorption calculated based on the assumption that the surface is covered with close-packed hemimicelles.

B. Polymer Adsorption

As mentioned in the introduction, the process of polymer adsorption is fairly complicated [38, 39]. This is due to the large number of possible conformations which a polymer can adopt at the solid/liquid interface. This was recognized as long as 1957 by Jenckel and Rumback [40] who found that the amount of polymer adsorbed per unit area of the surface would correspond to a layer several molecules thick if all segments of the chains are attached. They suggested a model in which each polymer is attached in sequences separated by loops that extend into solution. In other words, not all of the segments of a macromolecule are in contact with the surface. Sequences of segments that are in direct contact with the surface are termed "trains," those in between and extending to solution are termed "loops"; the free ends of the molecule also extending into solution are termed "tails." However, this picture, schematically shown in Figure 3.7a of a random sequence of tails, trains, and loops only applies to the case of a random homopolymer on a plane surface. Various other configurations may be distinguished, depending on the structure of the polymer and the nature of the groups involved. For example, if the chain contains groups with a preferential affinity to the surface, the configuration is no longer random since this group will form the adsorption point [31]. This is illustrated in Figure 3.7b whereby the preferentially adsorbed groups are in relatively short blocks. Clearly if all groups have a strong affinity for the surface, the polymer chain will adopt a flat configuration (Figure 3.7c) with most or all of the surface sites occupied. This situation is seldom realized in practice since both kinetic and entropic reasons dictate that a considerable fraction of the segments will be in loops or tails.

On the other hand, if the polymer molecule is formed from two blocks with large differences in their affinity to the surfaces, i.e., of the AB block type (with B having the larger affinity), the configuration shown in Figure 3.7d may result, whereby the B group adopts a loop–

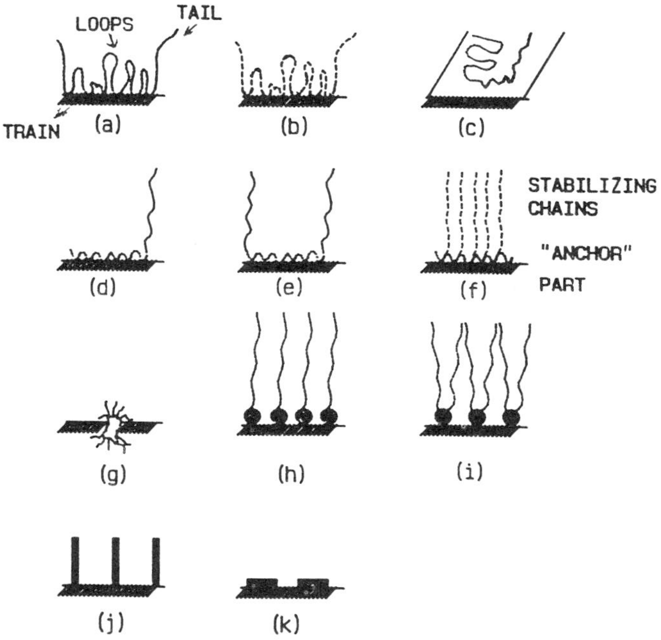

Figure 3.7 Various conformation of macromolecules adsorbed on a plane surface. (a) Random confirmation of loops–tails–tails–trains (homopolymer); (b) Preferential adsorption of short "blocks"; (c) Chain lying totally on the surface; (d) A block with loop–train configuration for B and one long tail for A; (e) ABA block as with (d); (f) BA_n graft with backbone B forming small loops leaving tails of A ("teeth"); (g) "Star"-shaped molecule; (h) Single-point attachment at chain end; (i) Single point attachment at middle of chain; (j) "Rod"-shaped molecule lying vertical; (k) Rod-shaped molecule lying horizontal.

train configuration (with small loops) leaving all the A segments dangling in solution as one long tail. A similar configuration is obtained with an ABA block (Figure 3.7e) except in this case two long dangling tails are present. A variation of this configuration is that obtained with a graft copolymer of the BA_n type (sometimes referred to as a comb-type structure), whereby now the B backbone adsorbs in the surface (either flat or with a "loopy" conformation) leaving a number of dangling tails (sometimes referred to as "teeth") in solution (Figure 3.7f). A number of structures are obtained with a star-shaped polymer whereby a number of tails originate from around a central block (Figure 3.7g). In certain cases, the whole configuration may be totally formed from tails (single point attachment; Figure 3.7h) as, for example, the case with a polymer molecule terminating with a functional group that is chemically bound to the surface. A slight variation of this (Figure 3.7i) is where the functional group is in the middle of the chain, whereby now two tails represent the configuration. Other types of configurations are obtained with rod-shaped molecules whereby these are either vertically (Figure 3.7j) or horizontally (Figure 3.7k) attached.

It is clear from the above description of polymer configuration that for full description of polymer adsorption it is necessary to obtain the following parameters (1) the amount of polymer adsorbed per unit area of the surface Γ; (2) the fraction of segments p that are in close contact with the surface, i.e., in trains and how strongly these are adsorbed; (3) the extension of the adsorbed polymer layer, i.e., the adsorbed layer thickness δ. It is necessary to know how these parameters change with the various parameters of the system such as surface coverage, molecular weight of the polymer, and solvency of the medium for the chains. A brief summary is given of how each of these parameters may be measured, showing some experimental results to illustrate the various trends obtained.

1. Amount of Polymer Adsorbed Γ (Adsorption Isotherms)

The determination of polymer adsorption, i.e., Γ, as a function of equilibrium concentration in solution C_2 is fairly well established. Basically one determines the change in polymer concentration ΔC, in

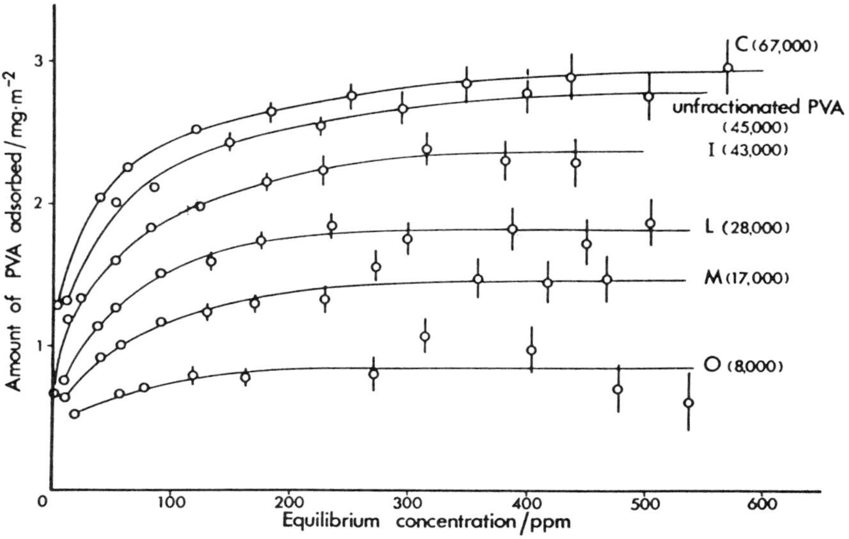

Figure 3.8 Adsorption isotherms of poly(vinyl alcohol) on polystyrene latex at 25°C.

bulk solution, after equilibration with the solid particles (of known surface area). It is essential to develop analytical techniques that are capable of measuring low concentrations (ppm) in order to establish the initially rising part of the isotherm, commonly found with polymers (high affinity isotherm). It is essential to wait for a long time to reach equilibrium since most polymers diffuse slowly in bulk solution and adsorption equilibrium may require several hours or even days. This is particularly the case at high coverage and where the polymer is polydisperse. The smaller molecules, which have higher diffusion coefficients, will adsorb first and these are gradually replaced by the preferentially adsorbed larger molecules.

As an illustration, Figure 3.8 shows the adsorption isotherms at 25°C for polyvinyl alcohol (PVA) on polystyrene latex (a typical hydrophobic surface) [41]. The polymer was a commercially available sample (Alcotex 88/10) that contains 12% acetate groups, i.e., it is

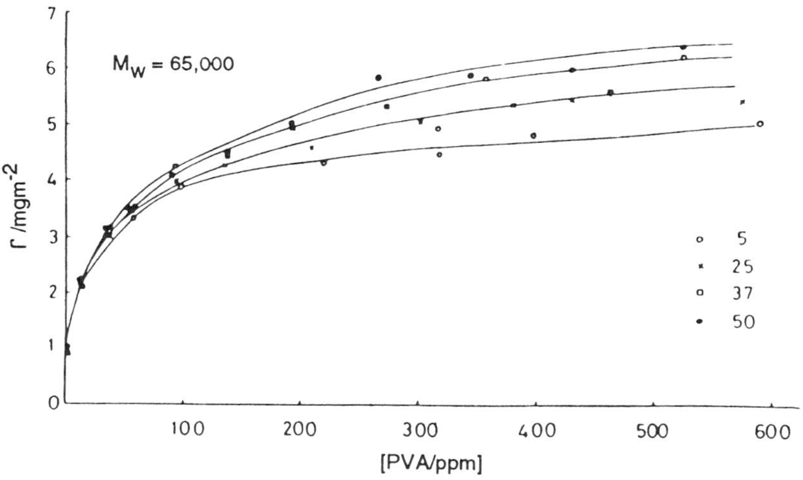

Figure 3.9 Adsorption isotherms for polyvinyl alcohol (M_w = 65,000) on polystyrene latex at various temperature.

88% hydrolyzed polyvinyl acetate. This polymer was fractionated using preparative gel permeation chromatography [41] or by a sequentional precipitation technique using acetone [42] and the molecular weight of the fractions was determined by ultracentrifugation and intrinsic viscosity measurements. The results of Figure 3.8 clearly illustrate the high affinity of the adsorption (i.e., adsorption is virtually irreversible) and that adsorption increases with increased molecular weight, as expected.

The influence of solvency of the medium on adsorption is illustrated in Figures 3.9 and 3.10, which show the adsorption as a function of temperature (PVA; M_w = 65,000) [42] and addition of KCl (for an unfractionated polymer) [43]. An increase in temperature or the addition of electrolyte reduces the solvency of the medium (water) for the PVA chains and in both cases adsorption increases as illustrated in Figures 3.9 and 3.10.

Results for polymer adsorption on agrochemicals are scarce. However, Tadros et al. [32, 43] showed similar trends for polymer adsorp-

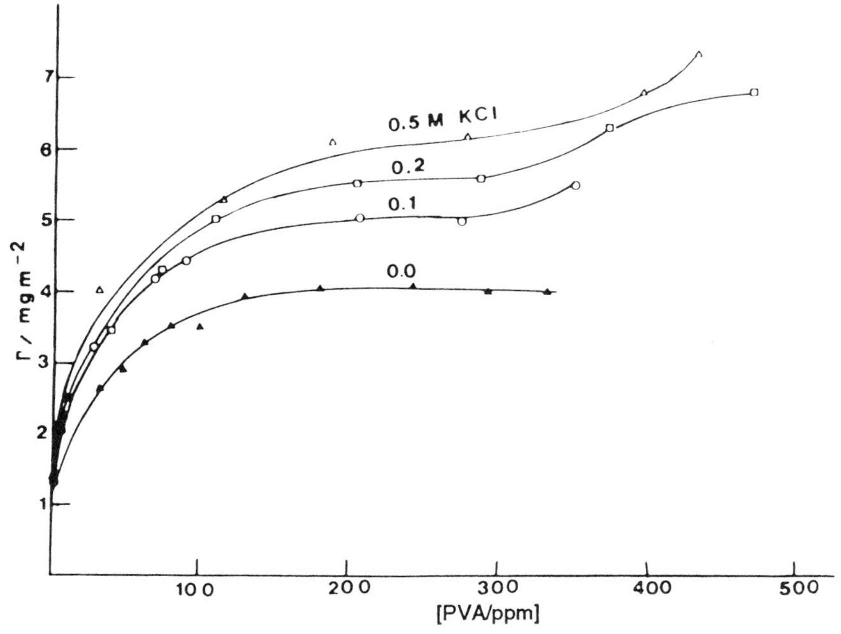

Figure 3.10 Adsorption isotherm for polyvinyl alcohol on polystyrene latex particles at various KCl concentrations.

tion on agrochemical particles. This is illustrated in Figures 3.11 and 3.12, which show the adsorption of PVA and a comb graft copolymer (polymethyl methacrylate backbone with polyethylene oxide side chains) on ethirimol (a fungicide) at room temperature. The high-affinity type of isotherm is clearly demonstrated and in both cases adsorption was irreversible, indicating high affinity to the surface. However, the amount of adsorption per unit area (using the BET surface area of ethirimol of 0.22 $m^2 g^{-1}$ obtained by Kr adsorption) was significantly higher than the values obtained on the model particles of polystyrene latex. This could be due to the errors involved in surface area determination of such coarse particles using BET gas, adsorption.

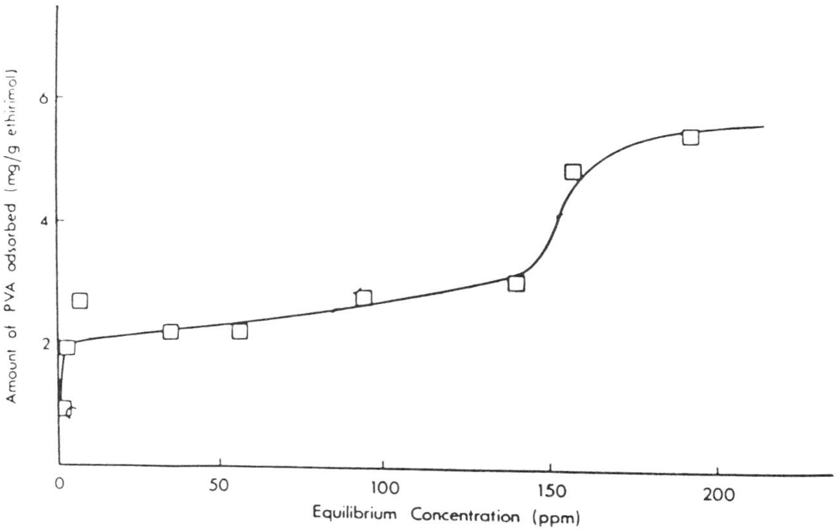

Figure 3.11 Adsorption of PVA on ethirimol.

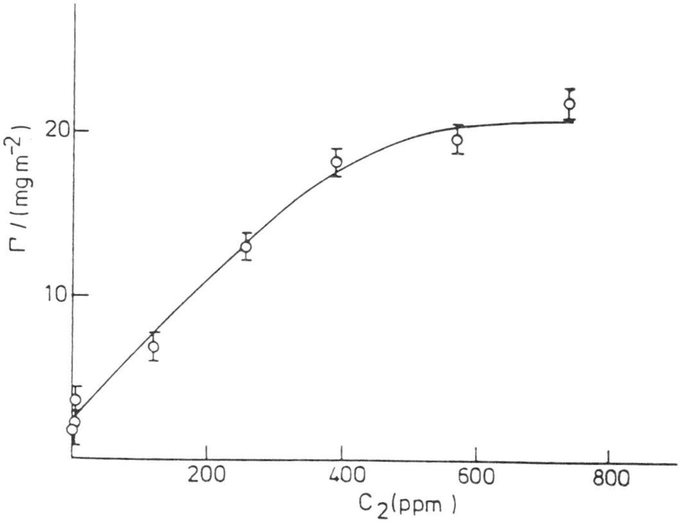

Figure 3.12 Adsorption of the "comb" graft on ethirimol surface.

2. Polymer Bound Fraction p

Several techniques are available for measurement of the fraction of segments of an adsorbed polymer that are in direct contact with the surface.

The most direct methods for assessing p are spectroscopic methods namely infrared (IR), electron spin resonance (ESR), and nuclear magentic resonance (NMR). The IR method depends on measuring the shift in some absorption peak for a polymer and/or surface group [44, 45]. ESR and NMR methods depend on the assumption that for an adsorbed homopolymer, the segments in trains have a lower mobility, i.e., a larger rotational correlation time than those in loops. In the ESR method [46], it is necessary to use a chemical spin label of a nitroxide group. The assumption is made that the introduction of a small number of such groups into the polymer does not affect the adsorption, particularly when such groups are randomly distributed. In this respect, the NMR method has a major advantage since it does not require attachment of a label and hence requires no such assumption. A pulsed NMR technique has been recently applied by Barnett et al. [47] for the estimate of p.

An indirect method for the estimation of p is microcalorimetry. Basically, one compares the enthalpy of adsorption per molecule with the enthalpy of adsorption per segment [48]. The ratio gives the number of segment per molecule in contact with the surface and hence one calculates p. The enthalpy of adsorption per segment is assumed to be equivalent to that of a small molecule of equivalent structure. The latter can be determined, for example, from measurement of the heat of wetting of such a small molecule on the same surface.

3. Adsorbed Layer Thickness

The practical methods for determination of the adsorbed layer thickness are mostly based on hydrodynamic techniques and hence the thickness is referred to as δ_h. Several measurements may be applied, of which viscosity, sedimentation coefficient and diffusion coefficient obtained from dynamic light scattering (photon correlation spectroscopy, or PCS) are the most common. A less accurate but easily accessible procedure is to compare the ζ potential of the particles with and without adsorbed layers. Below is given a brief summary of each of the above-mentioned techniques.

The viscosity method [49] depends on measuring the increase in the effective volume fraction of the particles as a result of the presence of an adsorbed layer thickness δ_h. Assuming that the particles behave as rigid noninteracting spheres, then the measured relative viscosity of the suspension η can be related to the effective volume fraction φ_{eff} (volume fraction of the particles φ plus the contribution from the adsorbed layer) by the classical Einstein equation, i.e.,

$$\eta_r = 1 + 2.5\varphi_{eff} \tag{53}$$

φ_{eff} and φ are related from simple geometry by the equation,

$$\varphi_{eff} = \varphi\left(1 + \frac{\delta_h}{R}\right)^3 \tag{54}$$

where R is the particle radius. Thus, from a knowledge of η_r and φ, one can obtain δ_h using Eqs. (53) and (54).

The sedimentation method depends on measuring the sedimentation coefficient, using an ultracentrifuge, of the particles (extrapolated to zero concentration), in the presence of the polymer layer [50]. Assuming that the particles obey Stokes' law, the sedimentation coefficient S_o' in the presence of an adsorbed layer is given by

$$S_o' = \frac{(4/3)\pi R^3(\rho - \rho_s) + (4/3)\pi\,[(R + \delta_h)^3 - R^3]\,(\rho_s^{ads} - \rho_s)}{6\pi\eta\,(R + \delta_h)} \tag{55}$$

where η is the viscosity of the medium, ρ and ρ_s are the mass density of the solid and solution phase, respectively, and ρ_s^{ads} is the average mass density of the interfacial region that may be obtained from the average mass concentration in the adsorbed layer.

A relatively simple sedimentation method for determination of δ_h that requires relatively cheap equipment is the slow-speed centrifugation technique applied by Garvey et al. [50]. Basically a stable monodisperse suspension is slowly centrifuged at low g values ($< 50g$) to form a close packed (hexagonal or cubic) lattice in the sediment. From a knowledge of packing modes (0.74 for hexagonal and 0.64 for random packing), the distance of separation between the centers of two particles R may be obtained, i.e.,

$$R_\delta = R + \delta_h$$

$$= \left(\frac{0.74\, V \rho_1 R^3}{W} \right)^{1/3} \tag{56}$$

where V is the sediment volume, ρ_1 the density of the particle, and W their weight.

The most rapid technique for measuring δ_h is photon correlation spectroscopy (PCS), which allows one to obtain the particle diffusion coefficient with and without adsorbed layers (D_δ and D, respectively). This is obtained from measurement of the intensity fluctuation of scattered light as the particles undergo Brownian motion [51]. When a light beam (e.g., monochromatic laser beam) passes through a dispersion, an oscillatory dipole measurement is induced in the particle, thus reradiating light. Due to the random arrangement of particles (which are separated by distances comparable to the wavelength of the light beam), the intensity of the scattered light will, at any instant, appear as random diffraction or "speckle" pattern. As the particles undergo Brownian motion, the random configuration of the speckle pattern changes. The intensity at any one point in the pattern will, therefore, fluctuate such that the time taken for an intensity maximum to become a minimum (i.e., the coherence time) corresponds approximately to the time required for a particle to move one wavelength. Using a photomultiplier of active area about the size of a diffraction maximum, i.e., approximately one coherence area, this intensity fluctuation can be measured. A digital correlator is used to measure the photo count or intensity correlation function of the scattered light. The photo count correlation function can be used to obtain the diffusion coefficient D of the particles. For monodisperse noninteracting particles, the normalized correlation function $[(g^{(1)}\ \tau)]$ of the scattered electric field is given by the equation:

$$[g^{(1)}(\tau)] = \exp - (\Gamma\ \tau) \tag{57}$$

where τ is the correlation delay time and Γ is the decay rate or inverse coherence time. Γ is related to D by the equation

$$\Gamma = D\,K^2 \tag{58}$$

where K is the magnitude of the scattering vector that is given by

$$K = \frac{4\pi n}{\lambda_o} \sin \frac{\theta}{2} \tag{59}$$

where n is the refractive index of the solution, λ_o is the wavelength of the light in vacuo, and θ is the scattering angle.

From D, the particles radius is calculated using the Stokes–Einstein's equation,

$$D = \frac{kT}{6\pi\eta R} \tag{60}$$

where k is the Boltzmann constant, and T the absolute temperature. For a polymer-coated particle R is denoted by R_δ which is equal to $R + \delta_h$. Thus, by measuring D_δ and D one can obtain δ_h. It should be mentioned that the accuracy of the PCS method depends on the ratio of (δ_h/R) since δ_h is obtained by difference. Since the accuracy of PCS is about 1%, δ_h should be at least 10% R. The same applies for the other above mentioned hydrodynamic method, which implies that one should choose small monodisperse particles. This means that application of these methods to agrochemical suspensions and emulsions is limited. Only indirect information is possible. One chooses a model spherical hydrophobic particle such as polystyrene latex that can be prepared monodisperse and obtain δ_h on these. The assumption is then made that δ_h will be of the same magnitude on the practical polydisperse agrochemical particles.

As mentioned above, electrophoretic mobility (u) measurements can also be used to measure δ_h [52]. This method can be applied to model and practical particles. From u, the ζ potential, i.e., the potential at the slipping plane of the particles, can be calculated. For a particle plus an adsorbed polymer layer, the thickness of the slipping plan is comparable to the hydrodynamic thickness δ_h. Assuming that the thickness of the Stern plane remains constant on the adsorption of a polymer, then the measured ζ potential in the presence of an adsorbed layer, ζ_d may be related to the Stern potential Ψ_d (which may be equated to the ζ potential of the bare particles) by the equation,

$$\tanh\left(\frac{e\zeta_\delta}{4kT}\right) = \tanh\left(\frac{e\zeta}{4kT}\right)\exp\left[-\kappa\left(\delta_h - \Delta\right)\right] \tag{61}$$

where κ is the Debye–Huckel parameter that is related to electrolyte concentration.

It should be mentioned that the value of δ_h calculated using the above simple Eq. (61) shows a dependence on electrolyte concentration and hence the method cannot be used in a straightforward manner. Recently, Cohen-Stuart et al. [53] showed that the measured electrophoretic thickness δ_e approaches δ_h only at low electrolyte concentration. Thus to obtain δ_h from electrophoresis, measurements should be obtained at various electrolyte concentrations and the results plotted vs. $(1/\kappa)$ to obtain the limiting value which now corresponds for δ_h.

4

Emulsifiable Concentrates

Many agrochemicals are formulated as emulsifiable concentrates (ECs) which when added to water produce oil-in-water (o/w) emulsions either spontaneously or by gentle agitation. Such formulations are produced by addition of surfactants to the pesticide if the latter is an oil with reasonably low viscosity or to an oil solution of the pesticide if the latter is a solid or a liquid with high viscosity. One of the earliest applications of ECs was reported by Jones and Fluno [54] who described an EC of dichlorodiphenyltrichloroethane (DDT). However, such formulations did not spontaneously emulsify on dilution in water [55] and were based on a single emulsifier. As we will see later, spontaneous emulsification requires a number of criteria that may be met by control of the properties of the interfacial region. Most of the progress on preparation of adequate ECs was based on a simple trial-and-error approach, in which a pair of emulsifiers was optimized for a specific agrochemical formulation. The first recommendation of the advantages of using a mixture of two or more emulsifiers was disclosed in the patent literature by Kaberg and Harris [56, 57] who used a mixture of sodium alkylbenzene sulfonate and ethoxylated nonionic surfactants. This composition was later slightly modified by using calcium or magnesium salts of alkylarylsulfonates as the anionic component of the surfactant blend [58]. With such blends, a 5% emulsifier concentration in the formulation may be sufficient for spontaneous emulsification and production of an emulsion with adequate stability within the time of application. Unfortunately, little fundamental work has been carried out to explain the reason for the good performance of such a blend. Indeed, most ECs are based on such mixture of anionic/nonionic surfactants and only slight modifications have been made to such a recipe. However, with the advent of agrochemical compounds, such simple blends were found, in some cases, to give inferior ECs. In addition, specific surfactants have to be developed to overcome some of the problems encountered with certain agrochemicals that may interact chemically with one or the two of the above-mentioned blend. Another problem that may be encountered with ECs is their sensitivity to variations in the batch of the chemical or the surfactants, which may result in lack of spontaneity of emulsification and/or the stability of the resulting emulsion. For this reason most manufacturers of ECs adopt rigorous quality control tests to ensure the adequacy of the resulting formulation under the practical conditions encountered in the field. This requires laborious testing of the effect of temperature, water hardness, agitation in the spray tank, and batch-to-batch variation of the ingredients of the formulation. As we will see later, the tests used by the formulation chemists are often too simple to provide adequate quantitative evaluation of the EC. Thus despite the wide use of ECs in agrochemicals, relatively little effort has been devoted to establish quantitative tests for assessment of the quality of the EC. A fundamental surface and colloid chemistry approach to the formulation of ECs is also lacking.

Due to the above shortcomings, a review on emulsifiable concentrates will at best be qualitative and will only help the reader in identifying the areas that require further research, rather than providing any principles or guidelines of how one can best formulate ECs. This contrasts the next two chapters on emulsions and suspension concentrates, whereby the basic principles are more established. A useful and recent review on ECs has been published by Becher [55] in this series of monographs. The present chapter will summarize some of the material already described by Becher [55]. The first part will deal with the common practice of formulation of ECs. This is then followed by a section on spontaneous emulsification, which is an important criterion for ECs. The question of stability of the resulting emulsion on standing will not be dealt in this chapter since this topic will be adequately covered in the next chapter on emulsions. The third section will deal with a specific example of an emulsifiable concentrate that has been investigated using fundamental studies.

I. GENERAL GUIDELINES FOR FORMULATION OF EMULSIFIABLE CONCENTRATES

As mentioned in the Introduction, ECs are formulated by a trial-and-error approach, whereby a pair of emulsifiers is selected for a specific agrochemical formulation. As stated by Becher [55], the hydrophilic–lipophilic balance (HLB) method that is normally used for selection of surfactants in emulsions (see Chapter 5) is inadequate for the formulation of ECs [59, 60]. This is not surprising since with ECs one requires in the first place spontaneity of dispersion on dilution, which as mentioned above is governed by the properties of the interfacial region. Other indices, such as the cohesive energy ratio concept suggested by Beerbower and Hill [61], may provide a better option. This concept will be discussed in some detail in Chapter 5. Essentially, the method involves selecting suitable emulsifiers by balancing the interactions of their hydrophobic parts with the oil phase and the hydrophilic parts with the aqueous phase. This involves knowledge of the solubility and hydrogen bond parameters of the various components. However, as will be seen, these parameters are not always available. Due to the lack of this information, it is not possible to check whether this approach could be applied to emulsifier selection.

From the above discussion, it is clear as to why the choice of surfactants for ECs is still made on a trial and error basis. Once this choice is made, extensive work is required to optimize the composition to produce an acceptable product that satisfies the criteria of spontaneous emulsification and stability under practical conditions. In many cases, it is also essential to add other components such as dyes, defoamers, crystal growth inhibitors, and various other stabilizers. The amount of work required for selection of emulsifiers may be illustrated from the publication of Kaertkemeyer and Ahmed [62] who investigated the emulsification of nonphytotoxic pesticidal oils using a set of eight nonionic surfactants. Four different hydrophobic groups—linear and branched alcohols (with an average of about 13 carbons) and linear and branched nonylphenol—were each ethoxylated to two different degrees to form the four pairs of surfactants investigated. In each case the oleophilic emulsifier was made by adding about 1 mol of ethylene oxide, while the hydrophilic one contained about 6 mol. The authors investigated the four combinations of two pairs possible when one pair had an alkyl hydrophobe and the other an alkylaryl one. The results were presented in the form of planes cutting the tetrahedral phase diagrams. Sections of the phase diagrams where good ECs existed were found, but they were only a small part of the total volume. No correlation was found between the HLB number of the surfactant blend and emulsion spontaneity or stability. The authors also noted that branching of the alkyl chains affected the results significantly. For example, the straight chain alkylphenols were found to be more effective than their branched counterparts. This shows the great sensitivity of the quality of the EC to minor changes in surfactant structure. This makes formulation of ECs somehow tedious, at least in some cases.

As mentioned above, testing of ECs is carried out using fairly simple procedures. The most common procedure is that based on the recommendation of the World Health Organization (WHO) which was originally designed for DDT emulsifiable concentrates. The WHO specification states that "any creaming of the emulsion at the top, or separation of sediment at the bottom, of a 100-ml cylinder shall not exceed 2 ml when the concentrate is tested as described in Annex 12 in *Specifications for Pesticides* [63, 64]. This test is described as follows. Into a 250-ml beaker having an internal diameter of 6–6.5 cm and 100-ml calibration mark and containing 75–80 ml of standard hard water, 5 ml of EC is added, using a Mohr-type pipette, during stirring with a glass rod, 4–6 mm in diameter, at about 4 revolutions per second (rps). The standard hard water is designed to contain 342 ppm, calculated as calcium carbonate. This is prepared by adding 0.304 g of anhydrous $CaCl_2$ and 0.139 g $MgCl_2 \cdot 6H_2O$ to make 1 liter using distilled water. The concentrate should be added at a rate of 25–30 ml/min, with the point of the pipette 2 cm inside the beaker, the flow of the concentrate being directed toward the center, and not against the side, of the beaker. The final emulsion is made to 100 ml with standard hard water, stirring continuously, and then immediately poured into a clean, dry 100-ml graduated cylinder. The emulsion is kept at 29–31°C for 1 hr and examined for any creaming or separation.

The reason for the above procedure is that both temperature and water hardness have a major effect on the performance of ECs. This is illustrated in Figures 4.1 and 4.2, which show the effect of water hardness on the amount of cream that separates from a typical formulation. The best performance appears to be observed at a water hardness of 300 ppm, but this may not be general with all other formulations. The rate at which the amount of cream approaches equilibrium is fairly independent of water hardness, which means that taking an arbitrary time of 1 hr to measure the separated cream or sediment is adequate for relative comparison of various formulations. Increasing the time to 3–4 hours for full creaming or sedimentation does not, in general, change the rating of various systems [65]. Figure 4.2 shows that the stability of the produced emulsion first improves as the water hardness is increased, but above a critical value of about 500 ppm it rapidly decreases with further increase in water hardness [66]. The decrease in the volume of cream at high water hardness is caused by the appearance of macroscopic oil droplets. At still higher water hardness the oil will separate as a distinct layer. Figure 4.3 shows the effect of temperature on the stability of an emulsion produced from an EC [67]. It is clear from this figure that the temperature variations encountered in practice have a significant effect on emulsion stability. Generally speaking, raising the temperature shifts the optimum performance to softer water and lowering it has the opposite effect.

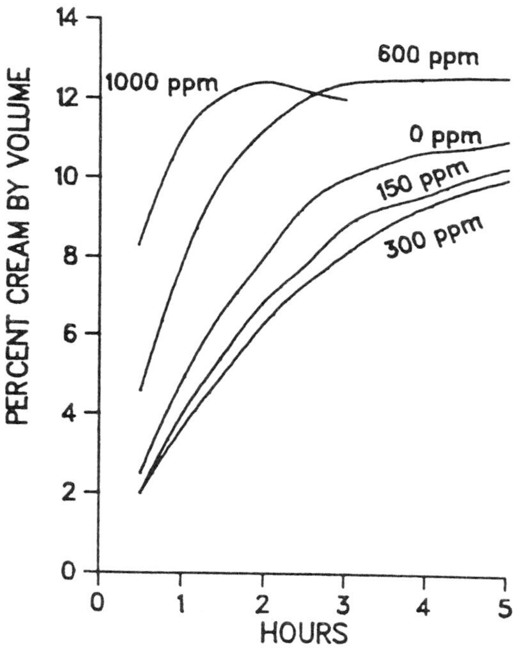

Figure 4.1 Effect of water hardness and time on the amount of cream that separates from an emulsion derived from a typical emulsifiable concentrate. (From Ref. 65.)

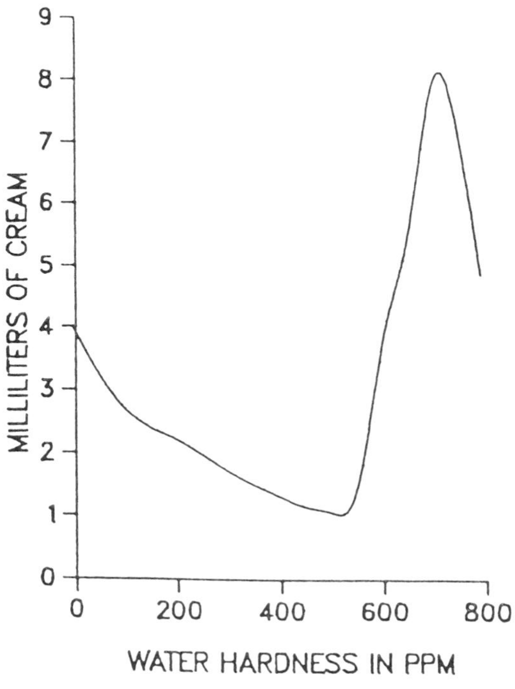

Figure 4.2 Effect of water hardness on emulsion stability. (From Ref. 66.)

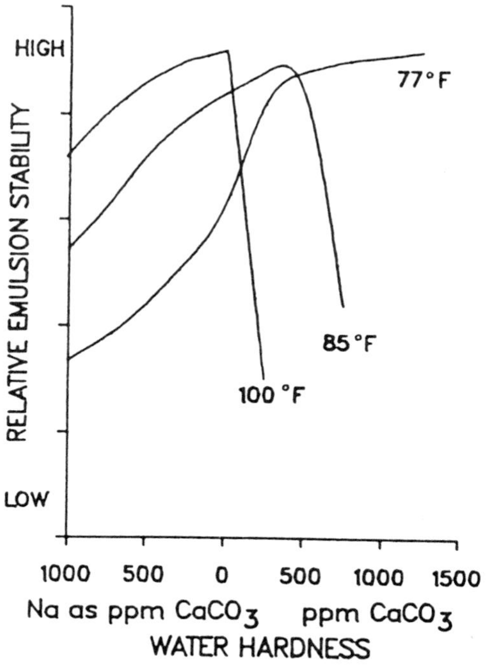

Figure 4.3 Effect of temperature on the stability of an emulsion derived from an emulsifiable concentrate. (From Ref. 67.)

The above dependence of the performance of ECs on water hardness and temperature may be related to the dependence of emulsifier properties on these parameters. This will be discussed in more detail in Chapter 5. The nonionic surfactants are particularly sensitive to these parameters. For example, the solubility in water and, therefore, the effective HLB number of a typical nonionic surfactant decrease as the temperature or salt content of the solution increases. This is evident from the decrease in cmc and cloud point with increasing salt concentration and/or temperature [68, 69].

Carino and Nagy [70] investigated the properties of ECs of toxaphene and diazinon dissolved in kerosene and xylene using surfactant

Table 4.1 Number of Moles of Ethylene Oxide that Gave the Most Stable Emulsions and the Number of Days Without Cream at Different Surfactant Ratios.

Moles of ethylene oxide			Number of days	
CaDBS (%)	Hard water	Soft water	Hard water	Soft water
33	9	8.5	< 1	< 1
50	12	11.5	6	6
60	17	15	41	27
67	21	19	52	26
71	26	23	24	14

blends of calcium dodecyl benzenesulfonate (CaDBS), (70%) and a series of ethoxylated nonylphenols. The results were compared in soft (water hardness of about 11.5 ppm) and hard (about 290 ppm) water. The number of ethylene oxide (EO) groups on the nonionic surfactant that gave the most stable emulsions and the number of days before any cream separated from them are given in Table 4.1 as a function of CaDBS concentration.

It can be seen from Table 4.1 that as the amount of CaDBS was increased, the number of EO units in the surfactant required to maintain stability also increased. The highest stability overall was found at a ratio of anionic to nonionic surfactant that was a function of water hardness. The authors also found a strong dependence of stability on the amount of surfactant used.

II. SPONTANEITY OF EMULSIFICATION

The first demonstration of spontaneous emulsification was demonstrated by Gad [71] who observed that when a solution of lauric acid in oil is carefully placed into an aqueous alkaline solution, an emulsion spontaneously forms at the interface. As we will explain later (Chapter

7), the reason for this spontaneous emulsification could be the very low (or transient negative) interfacial tension produced by the surfactant. Using an aqueous alkaline solution causes partial neutralization of lauric acid. A mixture of lauric acid and laurate can produce an ultralow interfacial tension. The process of spontaneous emulsification described by Gad [71] appears to occur with minimum external agitation, thus supporting the view that disruption of the interface may be occurring as a result of the combined surfactant film. Three main mechanisms may be responsible for spontaneous emulsification and these are briefly summarized below.

The first mechanism of spontaneous emulsification is due to interfacial turbulence that may occur as a result of mass transfer. In many cases the interface shows unsteady motions; streams of one phase are ejected and penetrate into the second phase, shedding small droplets. This is illustrated schematically in Figure 4.4 showing localized reductions in interfacial tension caused by the nonuniform adsorption of the surfactant at the oil–water interface [72] or by the mass transfer of surfactant molecules across the interface [73, 74]. With two phases that are not in chemical equilibrium, convection currents may be formed conveying liquid rich in surfactants toward areas of liquid deficient in surfactant [75, 76]. These convection currents may give rise to local fluctuations in interfacial tension, causing oscillation of the interface. Such disturbances may amplify themselves leading to violent interfacial perturbations and eventual disintegration of the interface, when liquid droplets of one phase are "thrown" into the other [77].

The second mechanism that may account for spontaneous emulsification is based on diffusion and stranding. This is best illustrated by carefully placing an ethanol–toluene mixture (containing, say, 10% alcohol) onto water. The aqueous layer eventually becomes turbid as a result of the presence of toluene droplets [78]. In this case interfacial turbulence does not occur although spontaneous emulsification apparently takes place. It has been suggested [79–80] that the alcohol molecules diffuse into the aqueous phase carrying some toluene in a saturated three-component subphase. At some distance from the interface the alcohol becomes sufficiently diluted in water to cause the

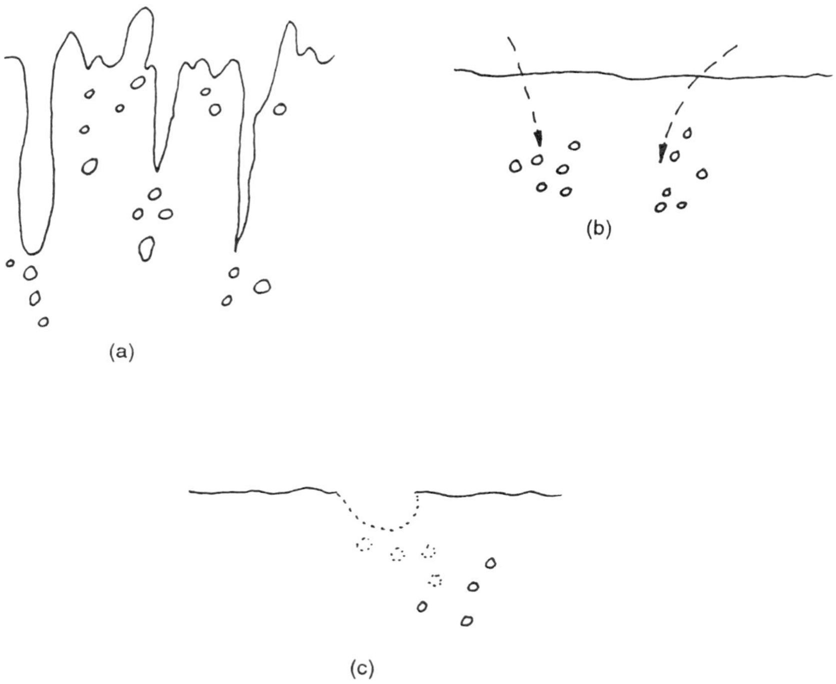

Figure 4.4 Schematic representation of spontaneous emulsification. (a) Interfacial turbulence, (b) diffusion and stranding, (c) ultralow interfacial tension.

toluene to precipitate as droplets in the aqueous phase. Such phase transition might be expected to occur when the third component increases the mutual solubility of the two previously immiscible phases.

The third mechanism of spontaneous emulsification may be due to the production of an ultralow (or transiently negative) interfacial tension. This mechanism is thought to be the cause of formation of microemulsions when two surfactants, one essentially water soluble and one essentially oil-soluble are used [81]. This mechanism will be described in detail in Chapter 7 on microemulsions.

III. FUNDAMENTAL INVESTIGATIONS ON A MODEL EMULSIFIABLE CONCENTRATE

Lee and Tadros [83–86] carried out some fundamental studies on a model EC of xylene containing a nonionic and a cationic surfactant. The objective of this work was to study the effect of stability of the resulting emulsion on herbicidal activity of a model compound, namely 2,4-dichlorophenoxyacetic acid ester. The nonionic surfactant used was Synperonic NPE 1800 (SNPE) (supplied by ICI), an ethoxylated-propoxylated nonylphenol having the following structure:

$$C_9H_{19}-C_6H_4-O-(\underset{CH_3}{\underset{|}{C}H}-CH_2-O-)_{13}-(CH_2-CH_2-O)_{27}H$$

The cationic surfactant was Ethoduomeen T20 (ET20) (supplied by Armour Hess) and has the following structure:

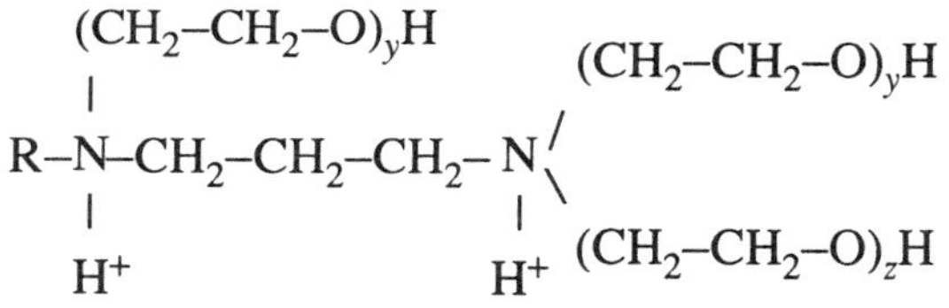

The cationic surfactant was used to ensure deposition of the resulting emulsion droplets on the negatively charged leaf surfaces. If some limited stability is induced in the resulting emulsion produced (by reducing the total surfactant concentration), the deposited emulsion droplets may undergo preferential coalescence at the leaf surface, thus enhancing contact with the herbicide and hence increasing biological efficacy.

The effect of surfactant concentration of SNPE + ET20 (in a 1:1 ratio by weight) on the spontaneity of dispersion on dilution of the xylene EC was studied using the CIPAC test and measuring the dispersed phase mean droplet diameter. In the CIPAC test 1 ml of an

EC was added by free fall from a 1 ml pipette held 1 cm above the surface to 100 ml of water contained in a 100 ml measuring cylinder. The apparent ease of dispersion was termed good, moderate, or poor. The measuring cylinder was then immediately inverted three times and the fineness and uniformity of the emulsion expressed on a scale of 0 (the sample does not emulsify) to 6 (the sample disperses to form a clear or opalescent solution, i.e., L_1, indicated on the phase diagram; see below). The average droplet diameter of each emulsion was measured immediately after dispersal of the EC using the Coulter Nanosizer. The instrument measures the time-dependent fluctuations in the intensity of scattered light by a dispersion, and calculates the average diffusion coefficient of the droplets and hence their average droplet diameter. Before the measurement, the emulsions were diluted in water to avoid multiple scattering. The results showed that a minimum of about 1% total surfactant is necessary to produce spontaneous emulsification, after which there is a gradual improvement in spontaneity and a reduction in droplet size with increase in surfactant concentration up to 5%. Beyond this concentration the average droplet size of the emulsions increases with increase in surfactant concentration, although the spontaneity of emulsification is well maintained until a concentration of 20% surfactant is reached. Any further increase in surfactant concentration results in deterioration of spontaneity accompanied by further increase in droplet size until, with 40% surfactant, the dispersal of the oil becomes relatively poor. When the concentration of surfactant reaches 60%, a very viscous EC is formed, which disperses very slowly to form a "solubilized" system with a small mean droplet diameter.

Figure 4.5 shows the change in viscosity of the ECs with increase in concentration of surfactant. This figure shows a rapid increase in viscosity above about 30% surfactant. This increase above 30% may be due to the formation of surfactant aggregates, e.g., inverse micelles produced in the presence of small amounts of water that are present in the hydrated surfactants.

Figure 4.6 shows the interfacial tension (γ) at the xylene–water interface vs. log concentration (expressed as %) for SNPE, ET20, and a 1:1 mixture of SNPE and ET20. It can be seen that SNPE is more

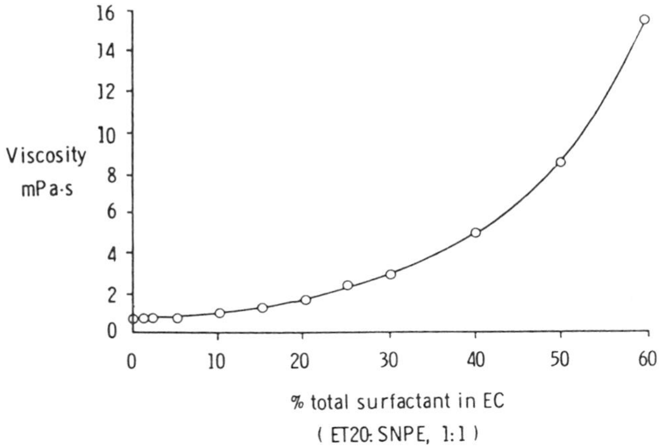

Figure 4.5 Variation of viscosity with surfactant concentration (1:1 mixture of SNPE and ET20) for a xylene EC.

surface active at the xylene–water interface than ET20. The γ values for the 1:1 mixture are closer to the values for SNPE, indicating preferential adsorption of SNPE at the oil–water interface. At sufficiently high concentration of surfactant the γ values become quite low (< 1 mN m^{-1}). This is clearly illustrated for the 1:1 mixture for which the concentration was extended to 2%. These low interfacial tensions were measured using the spinning drop technique [87]. No measurements could be made above 2% surfactant since the droplets disintegrated as soon as spinning of the tube started.

The above interfacial tension results may throw some light on the mechanism of spontaneous emulsification in the present model EC. As mentioned before, there are basically two main mechanisms of spontaneous emulsification: a creation of local supersaturation (i.e., diffusion and stranding) and mechanical breakup of the droplets as a result of interfacial turbulence and/or the creation of an ultralow (or transiently negative) interfacial tension. Diffusion and stranding is not the likely mechanism in the present system since no water-soluble

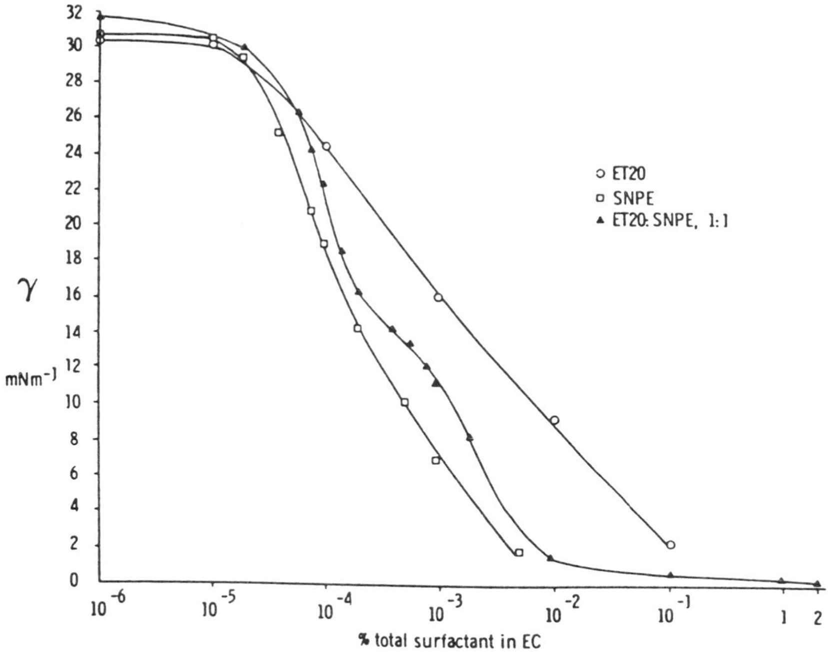

Figure 4.6 Interfacial tension: log concentration of surfactants.

cosolvent was added. To check whether the low interfacial tension produced is sufficient to cause spontaneous emulsification, a rough estimate may be made from consideration of the balance between the entropy of dispersion and the interfacial energy, i.e.,

$$\Delta G^{form} = \gamma\, dA - T\Delta S^{config} \tag{62}$$

where ΔG^{form} is the free energy of formation of the emulsion from the EC, dA is the increase in interfacial area when a bulk oil phase is dispersed into droplets, T is the absolute temperature, and ΔS^{config} is the configurational entropy of the droplets in the resulting dispersion. To a first approximation [88],

$$\Delta S^{config} = -nk\left[\ln \varphi_2 + \left(\frac{1-\varphi_2}{\varphi_2}\right)\ln(1-\varphi_2)\right] \tag{63}$$

where k is the Boltzmann constant and φ_2 is the volume fraction of the dispersed phase. It is clear that for spontaneous emulsification to occur, $\gamma\, dA < T\Delta S^{config}$. The limiting value of γ where this occurs is obtained by equating ΔG^{form} to zero. Replacing dA by $n\, 4\pi r^2$, where n is the number of droplets and r their radius), one obtains

$$\gamma = -kT\frac{[\ln \varphi_2 + (1-\varphi_2)/\varphi_2 \ln(1-\varphi_2)]}{4\pi r^2} \tag{64}$$

Taking an example from the present investigation, e.g., with 5% surfactant, $\varphi_2 = 0.01$ (the dilution used), and $r = 0.27$ μm, the value of γ required for spontaneous emulsification to occur is found from Eq. (64) to be about 2×10^{-5} mN m^{-1}. Values of this order have not been reached in the present investigations, thus ruling out the possibility that an ultralow interfacial tension is responsible for spontaneous emulsification. The most likely mechanism in the present system is interfacial turbulence that may be caused by mass transfer of surfactant molecules across the interface, which will also lead to interfacial tension gradients.

Another useful fundamental study for ECs is to establish the phase diagram of the various components. This is illustrated in Figure 4.7 for the present model EC of xylene/SNPE/ET20/water. The phase diagrams show the effect of addition of water on the three phase diagram of xylene/SNPE/ET20. The anhydrous ECs are all isotropic liquids (Figure 4.7a), with the exception of those containing SNPE concentrations in excess of its solubility in the other components, in which case solid SNPE is also present. At very high concentrations of SNPE a gel is formed. The ET20 is apparently miscible with the other components at all concentrations. The addition of 5% water to the ECs (Figure 4.7b) produces a large area of L_2 phase, comprising water solubilized with the inverse micelles of surfactant in xylene [89]. The phase diagram also shows small areas where an emulsion or an

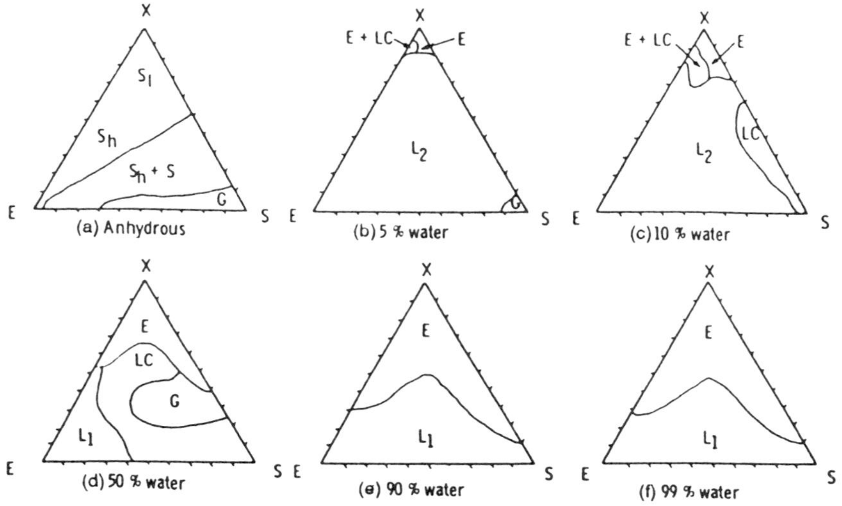

Figure 4.7 Phase diagrams for the system SNPE/ET20/xylene at various dilutions in water. X, xylene; E, ET20, S, SNPE; S_l, low viscosity solution; S_h, high viscosity solution; S, solid; G, gel; E, emulsion; LC, liquid crystal; L_2, organic isotropic solution; L_1, aqueous isotropic liquid.

emulsion in equilibrium with liquid crystalline phase is formed. This is particularly the case at low surfactant concentrations, where the water is not completely solubilized. With increasing concentrations of water (10, 50, 90, and 99% in Fig. 4.7c–f, respectively) the area of the emulsion phase increases, and inversion from water-in-oil to oil-in-water takes place at some unidentified concentration. With 10% and 50% water, a significant region of liquid crystalline phase is observed. With 50% water, a gel and L_1 phase (oil solubilized in an aqueous micellar solution) appear. With further increase in water concentration, the area of the L_1 phase extends at the expense of the liquid crystalline and gel phases. The latter phase disappears completely with 90% water leaving only emulsion and L_1 phases.

For the quantitative assessment of emulsion stability after dilution of the EC, it is necessary to measure the coalescence rate. This could

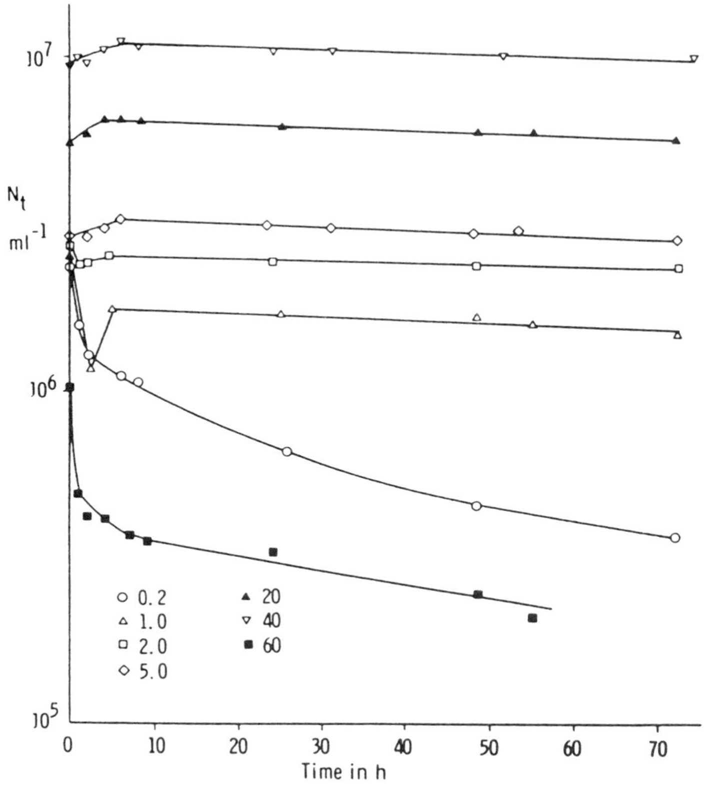

Figure 4.8 Log droplet number > 1 μm vs. time for spontaneously formed emulsions produced at various surfactant (1:1 ratio of SNPE/ET20 mixture) concentrations.

be done by measuring the droplet number as a function of time, using, for example, a Coulter counter. As an illustration, Figure 4.8 shows the results obtained using the model xylene EC. The number of droplets shown in Figure 4.8 are those greater than 1 μm, since the Coulter counter is not able to count submicrometer droplets. Assuming that coalescence occurs between aggregated oil droplets and that on average each droplet is in contact with two other droplets, then the rate of coalescence could be described by a first-order kinetics that is governed by the rupture of the aqueous film (lamella) separating neighboring droplets [90]. Thus if N_0 is the number of oil droplets at $t = 0$ and N_t is that at time t, then

$$N_t = N_0 \exp(-Kt) \qquad (65)$$

where K is the rate of coalescence. Equation (65) predicts that a plot of log N_t vs. t should be a straight line. However, straight lines were only obtained in very few cases (Fig. 4.8) and in the majority of cases the log N_t vs. t plots were curved showing some fluctuations in log N_t during the initial period (< 5 h). However, for the sake of comparison, the apparent coalescence rates were calculated from the slopes of the straight lines (when these were obtained) or of the tangents to the initial portion of each curve (within the first few hours). The coalescence rates of all emulsion were then plotted vs. total surfactant concentration, as shown in Figure 4.9. The later also shows the variation of the initial number of droplets > 1 μm vs. surfactant concentration.

The results in Figure 4.9 show that the coalescence rate K decreases very rapidly with increase in surfactant concentration from 0.2% to 1%, after which there is only a slight reduction in K with further increase in surfactant concentration until 40%. However, when the surfactant concentration is increased from 40% to 60%, the coalescence rate apparently increases to 2.14×10^{-4} s^{-1}, a value higher than that obtained with the lowest surfactant concentration (0.2%). There was also a simultaneous reduction in the initial number of droplets > 1 μm in diameter to 1.08×10^6 and the average droplet diameter as measured by the Coulter Nanosizer was 1.06 μm.

The increase in emulsion stability with increase in surfactant concentration up to 40% is what one would expect. This enhanced stability is due to an increase in the viscoelastic properties of the film, e.g., surface elasticity and/or viscosity and possible formation of liquid crystalline structures at the oil/water interface. This will be discussed in detail in Chapter 5. However, the increase in coalescence rate above 40% surfactant concentration is probably due to Ostwald ripening

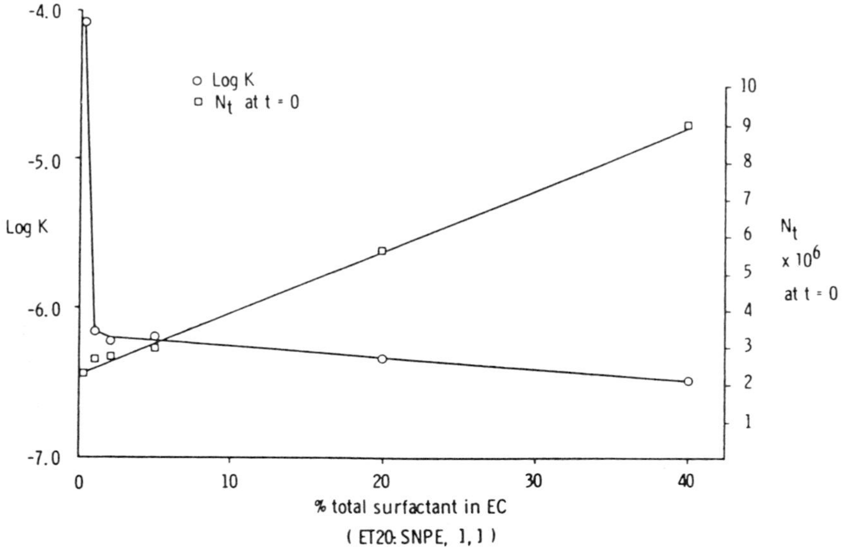

Figure 4.9 Coalescence rate and droplet number > 1 μm at $t = 0$ as a function of surfactant concentration.

which may be enhanced by solubilization by the surfactant micelles. The latter in particular is known to enhance crystal growth of solid/liquid dispersions [91]. The driving force for this process is the difference in solubility between small and large droplets. The smaller droplets with their greater solubility are thermodynamically unstable with respect to the larger ones. This can be expressed using the Ostwald–Freundlich equation [92]:

$$S_r = S_\infty \exp \frac{2\gamma M}{r\rho RT} \quad (66)$$

where S_r is the solubility of a droplet with radius r, S_∞ is the solubility of a droplet with infinite size, γ is the interfacial tension, M and ρ are the molecular weight and density of the droplet material, R is the gas constant, and T is the absolute temperature. Substitution of reasonable values of γ in Eq. (66) shows that the difference between S_r and S_∞ becomes significant as the droplet radius becomes very small. For example, taking a value of γ of about 1 mN m^{-1} (the value obtained at relatively high surfactant concentrations) S_r/S_∞ is 1.01 for a 0.01 μm droplet of xylene at 25°C. Such solubility difference does not in itself provide sufficient driving force to account for the instability of the emulsion at surfactant concentrations above 40%. Moreover, if the difference in solubility between large and small droplets is the only driving force accounting for instability, there is no reason for instability to continue to increase with increase in surfactant concentration in the range where the droplet size does not change significantly. Thus, some other process must be responsible for the transfer of the oil molecules from the smaller to the larger droplets. This is probably due to the solubilization of the oil by the surfactant micelles, an effect that is appreciable at high surfactant concentrations. The effect of this on droplet stability can be explained if one considers the diffusion of the oil molecules to the droplet/continuum interface. The diffusion flux, J, of the oil molecules in the aqueous continuous phase, in mol cm^{-2} s^{-1}, is given by Fick's first law:

$$J = -D\left(\frac{\delta c}{\delta x}\right) \tag{67}$$

where D is the diffusion coefficient and $\delta c/\delta x$ is the concentration gradient. As a result of solubilization, the oil molecules become incorporated into the micelles. Since the diffusion coefficient is roughly proportional to the radius of the diffusing particle [93, 94], D is reduced by a factor of about 10, which would correspond to a micelle having a volume 1000 times larger than that of the solubilizate. However, as a result of solubilization, the concentration gradient will increase greatly (in direct proportion to the extent of solubilization). This is because Fick's law involves the absolute gradient of concentration, which is small so long as the solubility is small, rather than its relative value. If S represents the saturation value, then [93, 94]:

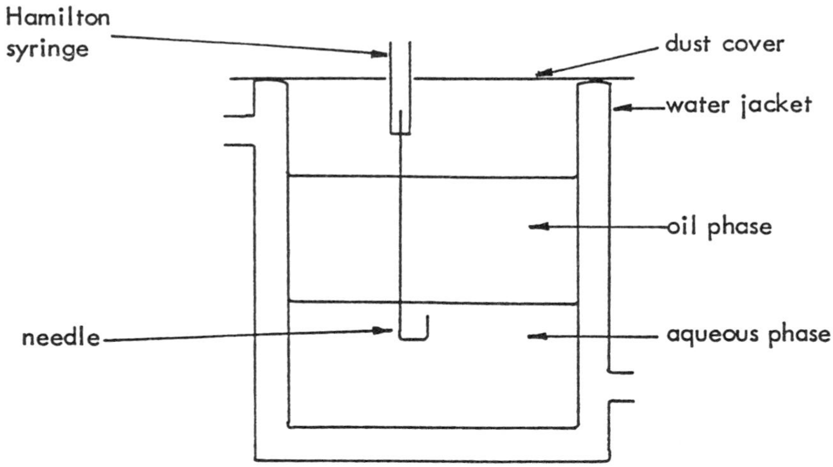

Figure 4.10 Setup for measuring coalescence at a planer oil/water interface.

$$J = -DS\left(\frac{\delta \ln S}{\delta x}\right) \tag{68}$$

Equation (68) shows that for the same gradient of relative saturation, the flux caused by diffusion is directly proportional to saturation. Hence solubilization will, in general, increase transport by diffusion, since it can increase the saturation value by many orders of magnitude, even though it decreases the diffusion coefficient. Thus, as a result of the large extent of solubilization at high surfactant concentrations, the diffusional flux increases and enhances the extent of Ostwald ripening. The greater the difference in size between the droplets, the greater the rate of growth of the larger droplets, particularly when there is a significant proportion of droplets in the submicrometer region. This was confirmed using homogenization after dilution of the emulsions to create smaller droplets. In this case Ostwald ripening was detected at much lower surfactant concentration, namely, 15% [84].

Another method that may be applied to investigate the stability of emulsions produced by dilution of ECs is to study the coalescence of droplets at a planer oil/water interface at various surfactant concentrations. This may be carried out using the following setup (Figure 4.10) suggested by Cockbain and McRoberts (95). Equal volumes of water and EC are placed in the cell, which is kept at constant temperature by circulating water from a thermostat bath through the double-walled vessel. The equal volumes of water and EC are left for several hours in the cell to equilibrate. Subsequently droplets of equilibrated EC, of the same volume, are individually formed from a Hamilton syringe in the aqueous phase near the interface and their rest times before coalescence with the bulk organic phase measured. Not less than 80 droplets should be measured when the rest times are long and more than 120 droplets when the rest times are short. As noted by many investigators [96–98], the rest times of a number of oil droplets produced from the same EC at constant surfactant concentration are not constant but show considerable variation. Two methods may be used to treat the data. In the first method, the rest times are assumed to be symmetrically distributed around a mean value (a Gaussian distribution) and an arithmatic mean (t_{mean}) and standard deviation (σ) are calculated for each system. The results obtained using this method are summarized in Table 4.2 for xylene ECs containing various concentrations of 1:1 w/w mixture of SNPE and ET20.

The second method used by Cockbain and McRoberts [95] involves plotting the results as a distribution curve. The number N_t of droplets that have not yet coalesced within a time t is plotted vs. time. This distribution curve consists of two fairly well-defined regions, one in which N_t is nearly constant with t, followed by a region in which N_t decreases with time in an exponential fashion. The first region corresponds to the process of drainage of the thin liquid film of the continuous phase from between the droplets and the planar interface, whereas the second region is that where rupture of the thin film and coalescence take place. Since film rupture and coalescence usually follow a first order process [95], the rate constant K can be calculated from the slope of the line of log N_t vs. t. As an illustration, the results obtained for xylene ECs containing 1:1 w/w mixture of SNPE:ET20 are shown in Figures 4.11 and 4.12 for various total surfactant concentrations. In these figures, log N_t/N_0 values are plotted vs. t, where

Table 4.2 Mean Rest Times, Half-Life for Film Rupture and Drainage Time at Various Surfactant Concentrations

% Total surfactant in EC	Gaussian distribution		Cockbain–McRoberts	
	t_{mean}/s	σ/s	$T_{1/2}$/s	t_D/s
0	5.1	3.2	4.5	0.6
1×10^{-4}	3.0	1.0	1.3	1.7
1×10^{-3}	1.3	0.3	0.5	0.8
2×10^{-3}	1.8	1.0	1.6	0.2
3×10^{-3}	7.0	3.8	7.4	(–0.4)
4×10^{-3}	14.0	10.6	19.0	(–5.0)
5×10^{-3}	98	40	41	57
6×10^{-3}	229	33	39	190
7×10^{-3}	240	19	38	202
1×10^{-2}	249	41	62	187
2×10^{-2}	272	27	48	224
3×10^{-2}	265	53	54	211
4×10^{-2}	259	36	47	212
5×10^{-2}	262	40	43	219
7.5×10^{-2}	267	31	37	230
1×10^{-1}	> 500	–	–	–
1.0	> 500	–	–	–

N_0 is the total number of droplets measured at $t = 0$. All these plots show the two regions mentioned above. A first-order half-life for film rupture ($T_{1/2}$) was calculated from the slope of the second section of each plot, using the relationship [95]:

$$\log \frac{N_t}{N_0} = -Kt \tag{69}$$

where K is the rate constant of film rupture that is equal to $\ln 2/T_{1/2}$. A value of drainage time t_D was also calculated from

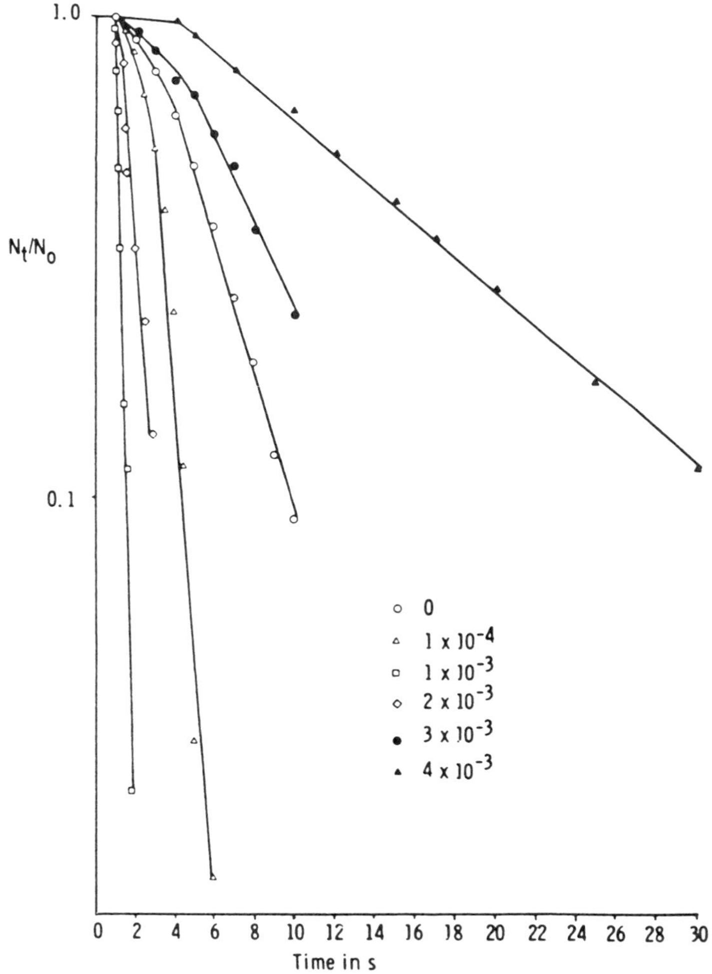

Figure 4.11 Cockbain–McRoberts plots for the ECs (0–0.004% surfactant).

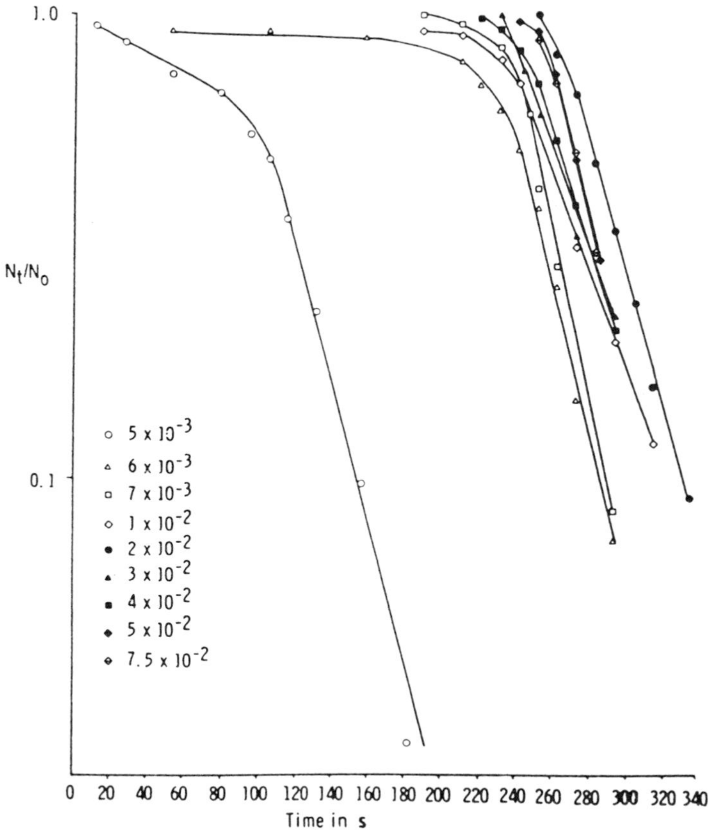

Figure 4.12 Cockbain–McRoberts plots for ECs (0.005–0.075% surfactant).

$$t_D = t_{mean} - T_{1/2} \tag{70}$$

where t_{mean} is the geometric mean rest time which is numerically equal to the experimental half-life $T_{1/2}$. t_D is a measure of the rate of drainage of the liquid film antecedent to its rupture.

The results for $T_{1/2}$ and t_D are summarized in Table 4.2. They show that the addition of surfactant to xylene results in an initial decrease in mean rest time and $T_{1/2}$, both of which reach a minimum value at $1 \times 10^{-3}\%$ total surfactant. With further increase in surfactant concentration, the mean rest time and $T_{1/2}$ increase, reaching their original values for xylene at approximately $3 \times 10^{-3}\%$ total surfactant. Above $3–4 \times 10^{-3}\%$ there is a sharp increase in drainage time and $T_{1/2}$ and an increase in the mean rest time.

The reduction in droplet rest time in the presence of small concentrations of surfactant ($< 4 \times 10^{-3}\%$) shown in Table 4.2 may be attributed to interfacial turbulence caused by the diffusion of surfactant molecules across the interface [96]. Such disturbances should increase the possibility of film rupture. However, this should not occur extensively in a chemically equilibrated system. The decrease in rest time at these low surfactant concentrations therefore indicates that either the systems were not fully equilibrated after having stood overnight or that interfacial turbulence was caused by changes in interfacial tension gradients caused by the disturbance of the surfactant film during drainage. Thus, one has to be careful in applying this technique for studying the stability of emulsions produced by dilution of the EC.

The sharp increase in drainage time and $T_{1/2}$ above $3–4 \times 10^{-3}\%$ surfactant implies the existence of a considerable resistance to film thinning (i.e., high emulsion stability) above this surfactant concentration. Several factors may account for such film stability and enhancement of the stability of the emulsion above this surfactant concentration: a reduction in interfacial tension, an increase in interfacial electrostatic potential, and/or the formation of a highly condensed interfacial film giving strong steric interactions between the droplet and the planar interface. For the sake of comparison, the mean rest time, interfacial tension, and ζ potential are plotted as a function

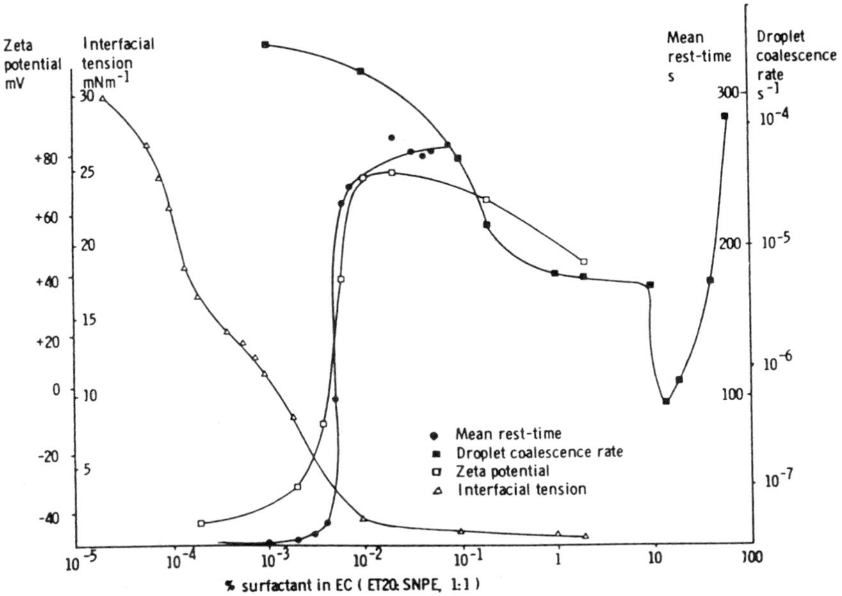

Figure 4.13 Mean rest times, interfacial tensions, ζ potentials, and coalescence rates as a function of surfactant concentration.

of surfactant concentration in Figure 4.13. The coalescence rates for the emulsions in bulk solution obtained by dilution of the ECs are also shown on the same diagram. It can be clearly seen that the increase in stability against drainage or coalescence is not directly related to reduction in interfacial tension, as the mean rest time did not significantly increase in the surfactant concentration range where the interfacial tension showed a sharp decrease. Indeed, it was not until the interfacial tension was significantly reduced below 5 mN m^{-1} that the rest time began to increase sharply. The same applies to the stability of the diluted emulsions. This behavior is not surprising since previous studies using nonionic [97] and anionic [98] surfactants showed little correlation between interfacial tension and resistance to drainage at a planar oil/water interface. However, the sharp increase in rest times seem to correlate with the sharp increase in ζ potential, although the correlation with coalescence rate of the diluted ECs is not as good.

It seems from the above discussion that a minimum surfactant concentration is required for ensuring the stability of the emulsion produced by dilution of the ECs. Mixed interfacial films with specific rheological properties are required for stabilization of the emulsions. These films should provide high dilational viscoelasticity and should prevent film thinning and drainage. This will be discussed in more detail in Chapter 5.

5

Emulsions

I. INTRODUCTION

Emulsions constitute a class of disperse systems consisting of two immiscible liquids one of which is the droplets (the disperse phase) and the second of which is the dispersion medium. The most common classes of emulsions are those whereby an oil is dispersed into water, referred to as oil-in-water (o/w) emulsions, or water dispersed into an oil, referred to as water-in-oil (w/o) emulsions. Clearly any two immiscible liquids can produce an emulsion, e.g., a polar oil such as ethylene glycol dispersed in a nonpolar oil such as a hydrocarbon. In this case, the emulsion may be referred to as an oil-in-oil (o/o) emulsion.

To disperse a liquid into another immiscible liquid, one usually requires a third component (see below), referred to as the emulsifier. It is perhaps useful to classify emulsions on the basis of the nature

Table 5.1 Classification of Emulsions

Nature of stability moiety	Structure of the system
Simple molecules and ions	Nature of the internal/external phase, e.g., o/w and w/o
Nonionic surfactants	Micellar emulsions (microemulsions)
Ionic surfactants	Macroemulsions
Surfactant mixtures	Multilayer droplets
Nonionic polymers	Double and multiple emulsions
Polyelectrolytes	Mixed emulsions
Biopolymers	
Mixed surfactants and polymers	
Liquid crystalline phases	
Solid particles (Pickering emulsions)	

of the stabilizing moieties or on the basis of the structure of the system. This is illustrated in Table 5.1 which reflects the complexity of the system [99].

Recently, many agrochemicals have been formulated as oil-in-water emulsion concentrates (ECs). These systems offer many advantages over the more traditionally used formulations of emulsifiable concentrates (described in Chapter 4). Being aqueous based, they may produce less hazard to the operator, e.g., they may cause less skin irritation. They may also be less phytotoxic to plants and they allow one to add water soluble adjuvants to the formulation. The emulsion systems can also be less expensive to produce since one replaces the oil used for the preparation of an EC, either completely or in part by an equivalent amount of water. The only disadvantage of emulsions vs. ECs is that they require special equipment for their preparation (such as high speed stirrers or homogenizers) and they may not be as physically stable as ECs.

In this chapter, I will discuss the basic principles involved in the preparation of emulsions and their stabilization. This is meant to be an introduction to the subject and for more details the reader should refer to the text recently edited by Becher [99].

II. FORMATION OF EMULSIONS

As mentioned in the Introduction, to produce an emulsion from two immiscible liquids one usually requires a third component, the emulsifier. The role of the emulsifier can be clearly understood from a consideration of the process of emulsification, which is schematically represented in Figure 5.1, whereby a bulk oil phase is subdivided into a large number of oil droplets [99]. Assuming the interfacial tension γ_{12} of the bulk oil and the droplets to be the same (this is usually true for droplets that are not too small, i.e., greater than say 0.1 μm), then the free energy of formation of the emulsion from the bulk phase is given by the simple expression,

$$\Delta G^{form} = \Delta A\ \gamma_{12} - T\Delta S^{config} \tag{71}$$

The first term $\Delta A\ \gamma_{12}$ is the energy required to expand the interface and

$$\Delta A = A_2 - A_1$$

is the increase in interfacial area. This term is positive since γ_{12} is positive. In the absence of an emulsifier, γ_{12} is of the order of 30–50 mN m^{-1} and hence $\Delta A\ \gamma_{12}$ is large and positive. Therefore, to reduce the energy for emulsification, one has to reduce γ_{12} by at least one order of magnitude. This is achieved by adsorption of the emulsifier at the o/w interface. As we will see later, the emulsifier will also play other roles to prevent breakdown of the emulsion by flocculation and coalescence. This is achieved by creating an energy barrier (due to electrostatic or steric repulsion) that prevents flocculation and an interfacial tension gradient (Gibbs elasticity) that prevents coalescence.

The second term in Eq. [71], $-T\,\Delta S^{config}$, is the configurational entropy resulting from the increase in number of possible configurations resulting from the production of a large number of droplets. This term, being negative, actually helps in formation of emulsions. However, with macroemulsions, $|\Delta A\ \gamma_{12}|$ is much larger than

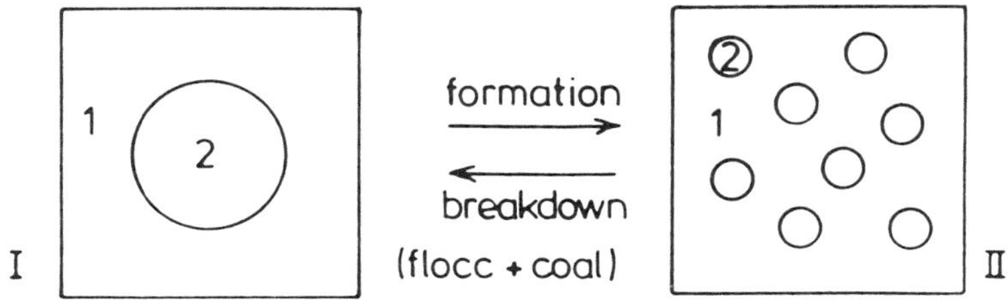

Figure 5.1 Schematic representation of emulsion formation and breakdown.

$|-T\,\Delta S^{config}|$ and hence ΔG^{form} is positive. In other words, emulsion formation is a nonspontaneous process and an energy barrier must be created to prevent the reverse to state I (by flocculation and coalescence). This means that emulsions are only stable in the kinetic sense and to give them a practical shelf life one has to maximize the energy barrier against flocculation and coalescence.

Several types of emulsifiers may be used to prepare an emulsion and these have been listed in Table 2.1. The emulsifier plays a number of roles in formation of the emulsion and its subsequent stabilization. The process of emulsification may be envisaged to start by formation of a film of the future (continuous) phase around the droplets. If no surfactant is present, this film is very unstable, draining rapidly under gravity, until complete drainage occurs. However, in the presence of a surfactant, the film can exist for some time as a result of the creation of an interfacial tension gradient $d\gamma/dz$. Such a gradient creates a tangential stress on the liquid or, alternatively, if the liquid streams along the interface with the surfactant, an interfacial tension gradient develops. This interfacial tension gradient supports the film, preventing its rupture by drainage (due to the gravitational force) providing

$$\frac{2d\gamma}{dz} > \rho_c h g$$

where h is the film thickness, ρ_c its density, and g the acceleration due to the gravity.

It should be mentioned, however, that the energy required for emulsification exceeds the thermodynamic energy $\Delta A\, \gamma_{12}$ by several orders of magnitude [100]. This is due to the fact that a significant amount of energy is needed to overcome the Laplace pressure, Δp, which results from the production of a highly curved interface (small droplets), i.e.,

$$\Delta p = \gamma\left(\frac{1}{R_1} + \frac{1}{R_2}\right) = \frac{2\gamma}{R} \qquad (72)$$

where R_1 and R_2 are the principal radii of curvature. For a spherical droplet with radius r,

$$\Delta p = \frac{2\gamma}{r}$$

Hence deformation leads to a large Δp and energy is needed to overcome this. This explains why emulsification is an inefficient process and why, to produce very small droplets, one needs to apply special methods, e.g., valve homogenizers, ultrasonics, static mixers, etc.

Five general main roles may be identified for the emulsifier. The first and most obvious is to lower γ, as mentioned above. This has a direct effect on droplet size; generally speaking, the lower the interfacial tension, the smaller the droplet size. This is the case when viscous forces are predominant, whereby the droplet diameter is proportional to γ. When turbulence prevails

$$d \propto \gamma^{3/5}$$

When emulsification continues, and an equilibrium is set up between the amount adsorbed and the concentration in the continuous phase,

C, the effective γ depends on the surface dilational modulus, ε, which is given by the equation

$$\varepsilon = \frac{d\gamma}{d \ln A} = A\left(\frac{d\gamma}{dA}\right) \tag{73}$$

where A is the area of interface (number of moles of surfactant adsorbed per unit area). Clearly ε depends on the nature of the surfactant. During emulsification, ε decreases as a result of depletion of surfactants and increase of $d \ln A/dt$. Hence the effective γ during breakup will be between the equilibrium value γ and γ_o (the interfacial tension of the bare liquid/liquid interface).

The second role of the surfactant is through its effect on the surface free energy for enlarging the drop surface. Both dilational elasticity and viscosity have an effect on the surface free energy for enlarging the drop surface. This surface free energy is now $\gamma dA + A\, d\gamma$, which implies that more energy is needed, although this energy is lower than $\gamma_o dA$. Moreover, if the surface dilational viscosity

$$\frac{d\gamma/d \ln A}{dt}$$

is large, viscous resistance to surface enlargements may cost entropy.

The third role of the surfactant is to create interfacial tension gradients. This has been discussed before. As a result of the tangential stress $d\gamma/dz$, which can build up on pressure of the order of 10^4 Pa (for $\gamma \sim 10$ mN m^{-1} and droplet diameter of 1 μm), the internal circulation in the droplet is impeded or even prevented, thus facilitating droplet formation and breakup.

The fourth role of the surfactant is to reduce coalescence during emulsification. The stabilizing mechanism of a surfactant during emulsification is usually ascribed to the Gibbs–Marangoni effect [101]. During emulsification, adsorption of surfactant is usually incomplete, so that the interfacial tension decreases with time and the film becomes rapidly depleted with surfactant as a result of its adsorption. The Gibbs elasticity, E_f, is given by the following equation [101–103]:

$$E_f = \frac{2\gamma\,(d \ln \Gamma)}{1 + (1/2)\,h\,(dC/d\Gamma)} \qquad (74)$$

where Γ is the surface excess (number of moles of surfactant adsorbed per unit area of the interface). As shown in Eq. (74), the Gibbs elasticity E_f will be highest in the thinnest part of the film. As a result, the surfactant will move in the direction of the highest γ and this motion will drag liquid along with it. The latter effect is the Marangoni effect. The final result is to reduce further thinning and hence coalescence is reduced. It should be mentioned that the Marangoni effect can be explained as liquid motion caused by the tangential stress $d\gamma/dz$. This gradient causes considerable streaming of liquid that forces its way into the gap between the approaching droplets, thus preventing their approach.

The fifth role of the surfactant is to initiate interfacial instability. Disruption of a plane interface may take place by turbulence, Rayleigh instabilities and Kelvin–Helmholtz instability. Turbulence eddies tend to disrupt the interface [104] since they create local pressures of the order of $(\rho_1 - \rho_2)\,u_e^2$ (where u_e is the shear stress velocity of the eddy, which may exceed the Laplace pressure $2\gamma/R$). The interface may be disrupted if the eddy size l_e is about twice R. However, disruption-turbulent eddies do not take place unless γ is very low. The Kelvin–Helmholtz instability arises when the two phases move with different velocities u_1 and u_2 parallel to the interface [105].

Interfacial instabilities may also occur for cylindrical threads of disperse phase that form during emulsification or when a liquid is injected into another from small orifices. Such cylinders undergo deformation [106, 107] and become unstable under certain conditions. This is illustrated in Figure 5.2, which shows that when there is a sinusoidal disruption of the radius of the cylinder, the latter becomes unstable when the wavelength λ of the perturbation exceeds the circumference of the undisturbed cylinder. Under these conditions, the

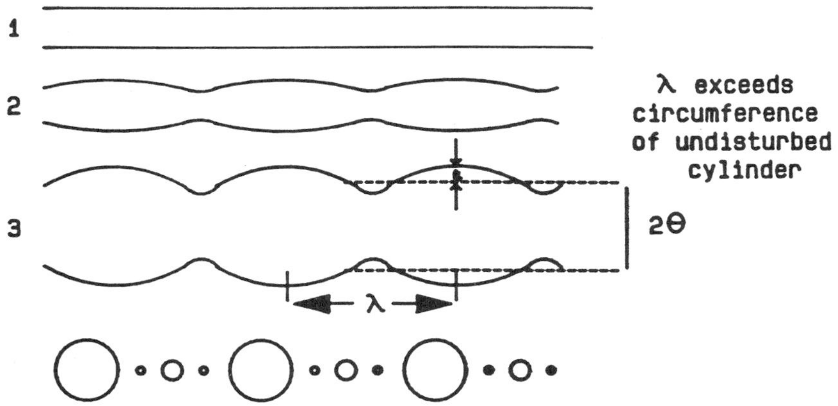

Figure 5.2 Schematic representation of disruption of a cylindrical thread of a liquid.

waves are amplified until the thread breaks up into droplets [108]. The presence of surfactants will accelerate this process of breakup due to interfacial tension gradients since the curved part will have a higher γ as a result of receiving the smallest amount of surfactant per unit area. Hence, surfactant is transported toward the point of strongest curvature, carrying liquid streaming to that part. This may cause droplet shredding. It is also clear from Figure 5.2 why emulsion formation leads to the production of polydisperse systems.

III. SELECTION OF EMULSIFIERS

The selection of different surfactants in the preparation of either o/w or w/o emulsion is still made on empirical base. One of the earliest semiempirical scales for selecting an appropriate surfactant or blend of surfactants was proposed by Griffin [109, 110] and is usually referred to as the hydrophilic–lipophilic balance or HLB number. This will be discussed below. Another closely related concept introduced by Shinoda and coworkers [111–113] is the phase inversion temperature (PIT) volume. Both the HLB and PIT concepts are fairly empirical and one should be careful in applying them in emulsifier selection. A more quantitative index that has drawn little attention is the cohesive energy ratio (CER), a concept introduced by Beerbower and Hill [114]. Below a brief description of the three indices is given.

A. The Hydrophilic–Lipophilic (HLB) Concept

This scale is based on the relative percentage of hydrophilic to lipophilic groups in the surfactant molecule(s). Surfactants with a low HLB number normally form w/o emulsions, whereas those with a high HLB number form o/w emulsions. A summary of the HLB range required for various purposes [110] is given in Table 5.2.

The relative importance of hydrophilic and lipophilic groups was first recognized when mixtures of surfactants were used with varying properties of surfactants having a low and high HLB number. The efficiency of any combination as judged by phase separation was found to pass through a maximum when the blend contained a particular concentration of the surfactant with the high HLB number. The original method for determining HLB numbers, developed by Griffin [109] is laborious and requires a number of trial and error procedures. Later, Griffin [110] developed a simple equation that permits calculation of the HLB number of certain numbers of nonionic surfactants such as fatty acid esters and alcohol ethoxylates of the type

Table 5.2 Summary of HLB Ranges and Other Applications

HLB range	Applications
3–6	w/o Emulsifier
7–9	Wetting agent
8–18	o/w Emulsifier
13–16	Detergent
15–18	Solubilizer

$R(CH_2{-}CH_2{-}O)_n{-}OH$. For the polyhydroxy fatty acid esters, the HLB number is given by the expression:

$$\text{HLB} = 20\left(1 - \frac{S}{A}\right) \tag{75}$$

where S is the saponification number of the ester and A the saponification number of the acid. Thus, a glyceryl monostearate, with S = 161 and A = 198, will have an HLB number of 2.8, i.e., it is suitable for a w/o emulsifier. However, in many cases, estimation of the saponification number accurately is difficult, e.g., ester of tall oil, resin, beeswax, and linolin. For the simple ethoxylate alcohol surfactants, the HLB can be simply calculated from the weight percent of oxyethylene E and polyhydric alcohol P, i.e.,

$$\text{HLB} = \frac{E+P}{5} \tag{76}$$

If the surfactant contains polyethylene oxide as the only hydrophilic group, e.g., in the primary alcohol ethoxylates $R(CH_2{-}CH_2{-}O)_n{-}OH$, the HLB number is simply ($E/5$) (the content from one OH group is simply neglected).

The above equation cannot be used for nonionic surfactants containing propylene oxide or butylene oxide, nor can it be used for ionic surfactants. In the latter case, the ionization of the head groups tends to make them even more hydrophilic in character, so that the HLB number cannot be calculated from the weight percent of the ionic groups. In that case, the laborious procedure suggested by Griffin [109] must be used.

Davies [115] derived a method for calculating the HLB number of surfactants directly from their chemical formulas, using empirically determined group numbers. Thus, a group number is assigned to various emulsifier component groups and the HLB number is then calculated from these numbers using the following empirical relationship [116],

Table 5.3 HLB Group Numbers

Group	Group number
Hydrophilic	
$-SO_4^-Na^+$	30.7
$-COO^-H^+$	21.2
$-COO^-Na^+$	19.1
N (tertiary amine)	9.4
Ester (sorbitan ring)	6.8
Ester (free)	2.4
-COOH	2.1
-O-	1.3
CH-sorbitan ring	0.5
Lipophilic	
-CH-	0.475
$-CH_2-$	
$-CH_3$	
Derived	
$-CH_2-CH_2-O$	0.33
$-CH_2-CH_2-CH_2-O$	−0.15

$$\text{HLB} = 7 + \sum(\text{hydrophilic group nos.}) - \sum(\text{lipophilic group nos.}) \quad (77)$$

The group numbers for various component groups are given in Table 5.3. Davies [115] has shown that the agreement between HLB numbers calculated using the above empirical equation and those determined experimentally is quite satisfactory.

Various procedures were later devised to determine the HLB number of different surfactants. For example, Griffin [114] found a good correlation between the cloud point of 5% solution of various nonionic surfactants and their HLB number. This enables one to obtain the

HLB number from a simple measurement of the cloud point. A more accurate method of determination of HLB number is based on gas–liquid chromatography [117].

B. The Phase Inversion Temperature (PIT) Concept

Shinoda and coworkers [111–113] found that many oil-in-water emulsions, when stabilized with nonionic surfactants, undergo a process of inversion at a critical temperature (PIT). They also showed that the PIT is influenced by the surfactant HLB number. Several conclusions were drawn from this work. First, the size of the emulsion droplets was found to depend on the temperature and HLB of the emulsifier. Second, the droplets are less stable toward coalescence close to the PIT. Third, relatively stable o/w emulsions were obtained when the PIT of the system was 20–60°C higher than the storage temperature. Fourth, a stable emulsion was obtained by rapid cooling of an emulsion that was prepared at the PIT. Finally, the optimum stability of an emulsion was found to be relatively insensitive to changes in HLB or PIT of the emulsifier but instability was very sensitive to the PIT of the system. Shinoda et al. [118] found that stability against coalescence increased markedly as the molar mass of the lipophilic and hydrophilic groups increase. They also found that maximum stability occurs when the distribution of the hydrophilic groups was broad. In the latter case, the cloud point is lower and the PIT is higher than in the corresponding case for narrow size distribution. Thus, the PIT and HLB number are directly related parameters and indeed a close correlation was found between the two numbers. In view of this correlation, Sherman and coworkers [119–121] suggested the use of measurements of PIT value as a rapid method of assessing emulsion stability. However, one should be careful in using these concepts since they are only applicable to limited types of surfactants and oils.

C. The Cohesive Energy Ratio Concept

Beerbower and Hill [114] considered the dispersing tendencies of the oil and water interfaces of the surfactant or emulsion region in terms of the cohesive energies of the mixtures of oil with the lipophilic portion of the surfactant and the water with the hydrophilic portion. They used the Winsor [122] R concept which is the ratio of intermolecular attraction of oil molecules and lipophilic portion of surfactant to that of water and hydrophilic portion.

Several interaction parameters may be identified at the oil and water sides of the interface. Representing the lipophilic portion by L, the oil by O, the hydrophilic portion by H, and the water by W, one can identify at least nine interaction parameters: C_{LL}, C_{OO}, C_{LO} (at the oil side), C_{HH}, C_{WW}, C_{HW} (at the water side) and C_{LW}, C_{HO}, C_{LH} (at the interface). In the absence of the emulsifier, there will only be three interaction parameters, i.e., C_{OO}, C_{WW} and C_{WO}. If $C_{OW} \ll C_{WW}$ the emulsion breaks. These interaction parameters may be related to the Hildebrand solubility parameter δ (at the oil side of the interface [123] and the Hansen [124] polar hydrogen bonding and polar contribution to δ at the water side of the interface. δ of any component is related to the heat of vaporization ΔH by the expression:

$$\delta^2 = \frac{\Delta H - RT}{V_M} \tag{78}$$

where V_M is the molar volume.

Hansen [124] considered δ to consist of three main contributions, a dispersion contribution δ_d, a polar contribution δ_p, and a hydrogen bonding contribution δ_h. These contributions have different weighting factors such that

$$\delta^2 = \delta_d^2 + 0.25\delta_p^2 + 0.25\delta_h^2 \tag{79}$$

Beerhower and Hills [114] used the following expression for the HLB number:

$$\text{HLB} = \frac{20\, M_H}{M_L + M_H}$$

$$= \frac{20\, V_H\, \rho_H}{V_L\, \rho_L + V_H\, \rho_H} \tag{80}$$

where M_H and M_L are the molecular weights of the hydrophilic and lipophilic positions of the emulsifier. V_H and V_L are their corresponding molar volumes whereas ρ_H and ρ_L are the densities, respectively.

The cohesive energy ratio was originally defined by Winsor [122] as

$$R_o = \frac{C_{LO}}{C_{HW}}$$

When

$$C_{LO} > C_{HW},\ R_o > 1$$

and a w/o emulsion forms; if

$$C_{LO} < C_{HW}\ ,\ R_o < 1$$

and an o/w emulsion forms, whereas if

$$C_{LO} = C_{HW},\ R_o = 1$$

and a planer system results. The last case denotes the inversion point. R_o can be related to V_L, ρ_L, V_H and ρ_H by the expression,

$$R_o = \frac{V_L\, \rho_L^2}{V_H\, \rho_H^2}$$

$$= \frac{V_L\, (\delta_d^2 + 0.25\delta_p^2 + 0.25\delta_h^2)_L}{V_H\, (\delta_d^2 + 0.25\delta_p^2 + 0.25\delta_h^2)_H} \tag{81}$$

Combining Eqs. (80) and (81) one obtains the following general expression for the cohesive energy ratio:

$$R_o = \left(\frac{20}{\mathrm{HLB}} - 1\right) \frac{V_L\,(\delta_d^2 + 0.25\delta_p^2 + 0.25\delta_h^2)_L}{V_H\,(\delta_d^2 + 0.25\delta_p^2 + 0.25\delta_h^2)_H} \tag{82}$$

Thus, for an o/w emulsion, HLB = 12–15 and R_o = 0.58–0.29 ($R < 1$). For a w/o system, HLB = 5–6 and R_o = 2.30–1.90 ($R > 1$), whereas for a planer system, HLB = 9–10 and R_o = 1.25–0.85 ($R \sim 1$). Thus R_o combines both HLB and cohesive energy densities, giving a more quantitative estimate for emulsifier selection. The applicability of this concept relies on the availability of data for δ_d, δ_p, and δ_h for the various surfactant portions. Some values have been tabulated by Beerbower and Hills [114] and later by Barton [125]. Clearly, a comprehensive list of the group contributions for various surfactants is still unavailable and some measurements and calculations are needed before such a concept can be applied in practice.

IV. EMULSION STABILITY

Various emulsion breakdown processes may be identified and these are schematically represented in Figure 5.3. These breakdown processes will be briefly summarized below and particular attention will be paid to how one can stabilize the emulsion against the particular instability described.

A. Creaming and Sedimentation

Creaming and sedimentation processes result from external forces, usually gravitational or centrifugal. When such forces exceed the thermal motion of the droplets (Brownian motion) a concentration gradient builds up in the system with the larger droplets moving faster to the top (if the density is lower than that of the medium) or to the bottom (if the density is larger than that of the medium) of the container. In limiting cases, the droplets may form a close-packed (random or

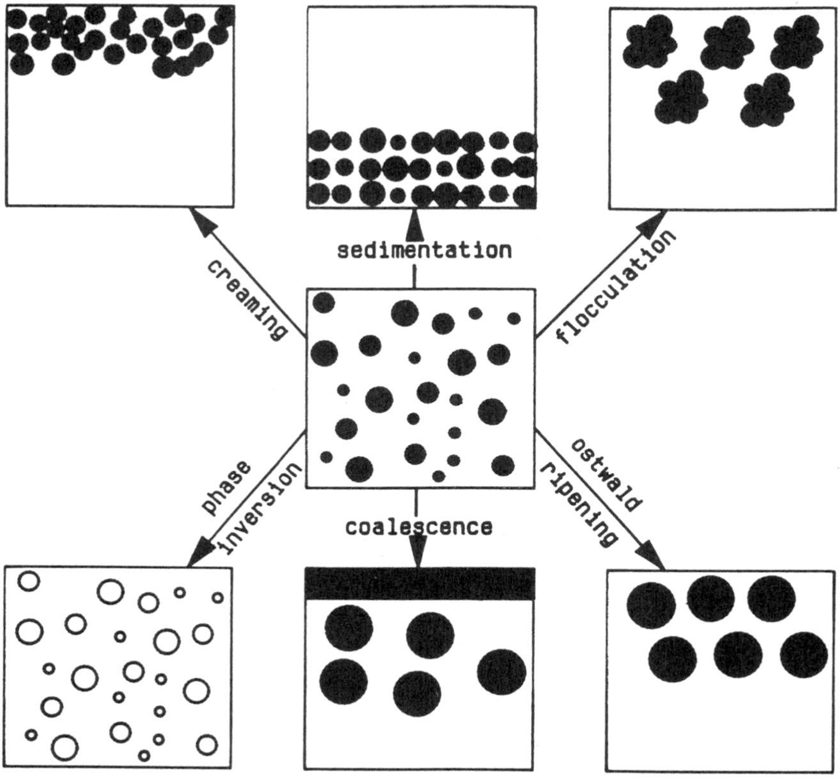

Figure 5.3 Schematic representation of emulsion breakdown processes.

ordered) array at the top or bottom of the system with the remainder of the volume being occupied by the continuous phase liquid. The case where the droplets move to the top is referred to as creaming whereas that whereby the droplets move to the bottom is referred to as sedimentation. Strictly speaking, when one refers to creaming or sedimentation, it is understood that no change in droplet size or distribution takes place.

As mentioned, the process of creaming or sedimentation is opposed by the thermal (Brownian) motion of the droplets and since this force increases with decrease in droplet size, then both processes are significantly reduced. Indeed, for no separation to occur, the Brownian diffusion kT (where k is the Boltzmann constant and T the absolute temperature) must exceed the gravitational force, i.e.,

$$kT > \frac{4}{3} \pi R^3 \Delta\rho \, g \, L \tag{83}$$

where $\Delta\rho$ is the density difference between dispersed phase and medium, and L is the height of the container. It is clear that to reduce creaming or sedimentation, $\Delta\rho$ has to be made close to zero (i.e., the density of the oil should be as close as possible to that of the medium) and R as small as possible. This simply follows from Stokes's law which gives the sedimentation velocity for a very dilute emulsion consisting of noninteracting droplets, i.e.,

$$v_o = \frac{2 R^2 \Delta\rho \, g}{9 \eta_o} \tag{84}$$

Equation (84) only applies for an infinitely dilute emulsion. For a concentrated emulsion, the creaming or sedimentation rate v is reduced with increase in the volume fraction of the emulsion. This can be empirically expressed as

$$v = v_o (1 - k\varphi) \tag{85}$$

where k is an empirical constant that accounts for droplet–droplet interaction. For the simplest case, where hydrodynamic interaction is only considered, k is in the region of 5–6. This is usually the case at $\varphi < 0.1$. However, for more concentrated emulsions k becomes a complex function of φ. Usually v decreases with increase in φ, approaching zero, as φ approaches φ_p, the so-called maximum packing fraction. φ_p is in the region of 0.7 for fairly monodisperse emulsion, but it can reach higher value (> 0.8) for polydisperse systems. However, when one approaches the maximum packing fraction, the viscosity of the emulsion becomes very high, (at $\varphi = \varphi_p$, $\eta \sim \infty$). Therefore, most practical emulsions have volume fractions well below φ_p and creaming or sedimentation is the rule rather than the exception. In this case, various procedures must be applied to avoid emulsion separation. As mentioned above, one of the simplest methods is to reduce $\Delta\rho$ or R. This may be achieved by matching the density of the oil to that of the medium (by using oil mixture) and/or reducing R by the use of homogenizers. In cases where this is not possible in practice one may use thickeners or apply the concept of controlled flocculation. Thickeners are perhaps the most widely used materials for reducing creaming or sedimentation. These are usually high molecular weight polymers of the synthetic or natural type, e.g., hydroxyethylcellulose, polyethylene oxide, xanthan gum, guar gum, alginates, carrageenans, etc. All these materials when dissolved in the continuous phase increase the viscosity of the medium and hence reduce creaming or sedimentation. However, their action is not simple, since these materials give non-Newtonian solutions that are viscoelastic. This means that the viscosity of these polymer solutions depends on the applied shear rate $\dot{\gamma}$. Generally speaking such systems produce pseudoplastic flow with an apparent yield value τ_β (the stress extrapolated to $\dot{\gamma} = 0$) and an apparent viscosity η_{app} which decreases with increase of $\dot{\gamma}$. Moreover, the viscosity of these polymer solutions increases with increased concentration in a peculiar way. Initially η increases with C, but above a certain concentration, to be denoted C^*, there is a much more rapid increase in η with further increase in C. This concentration C^* denotes the point at which polymer coil overlap begins to occur and above C^*, the solution shows elastic behavior which increases with increase in C. Usually, thickeners are added at concentrations that are above C^*, in which case a viscoelastic system is produced. Moreover, C^* decreases with increase in molecular weight of the added polymer. Hence to reduce the polymer concentration above which a viscoelastic system is produced, one uses higher molecular weights.

The above-mentioned viscoelastic polymer solutions reduce (or eliminate) creaming or sedimentation of the emulsion, providing they produce an "elastic" network in the continuous phase that is sufficient to overcome the stresses exerted by the creaming or sedimenting droplets. Such viscoelastic solutions produce a very high zero shear viscosity that is sufficient to eliminate creaming or sedimentation.

Another method of reducing creaming or sedimentation is to induce weak flocculation in the emulsion system. This may be achieved by controlling some parameters of the system such as electrolyte concentration, adsorbed layer thickened, and droplet size. These weakly flocculated emulsions will be discussed in the next section. Alternatively, weak flocculation may be produced by addition of a "free" (nonadsorbing) polymer. Above a critical concentration of the added polymer, polymer–polymer interaction becomes favorable as a result of polymer coil overlap and the polymer chains become "squeezed out" from between the droplets. This results in a polymer free zone between the droplets. Weak attraction occurs as a result of the higher osmotic pressure of the polymer solution outside the droplets. This phenomenon is usually referred to as depletion flocculation [126] and could be applied for "structuring" emulsions and hence reduction of creaming or sedimentation.

B. Flocculation of Emulsions

Flocculation of emulsions refers to aggregation of the droplets (see Figure 5.3), without any change in the primary droplet size, into larger units. Flocculation is the result of van der Waals attraction, which is universal for all disperse system. For two droplets of equal radii R, the van der Waals attractive forces V_A is given by the following expression [127] (when the distance of separation between the droplets h is much smaller than the droplet radius):

$$V_A = -\frac{A\,R}{12\,h} \tag{86}$$

where A is the Hamaker constant, between droplets A_{11} and medium A_{22}, i.e.,

$$A = (A_{11}^{1/2} - A_{22}^{1/2})^2 \tag{87}$$

The Hamaker constant of any material is given as

$$A_{ii} = \pi q^2 \beta_{ii} \tag{88}$$

where q is the number of atoms or molecules per unit volume and β_{ii} is the London dispersion constant (which is related to the polarizability).

It is clear from Eq. (86) that V_A increases rapidly with decreasing distance of separation between the droplets. In the absence of any repulsion between the droplets, flocculation is very fast producing clusters of droplets. Thus, for stabilization of droplets against aggregation, a repulsive force must be created to prevent close approach of the droplets. Two general stabilizing mechanisms may be envisaged. The first is based on the creation of an electrical double layer around the droplets. This may be produced, for example, by adsorption of an ionic surfactant. In this case, the surface of the droplets becomes covered with a layer of charged head groups (negative with anionic and positive with cationic surfactants). This charge becomes compensated by counterions, some of which approach the surface closely (in

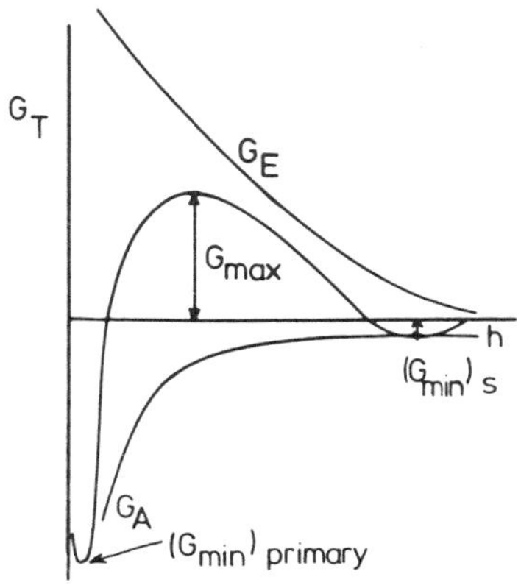

Figure 5.4 Schematic representation of energy–distance curve according to the DLVO theory [128, 129].

the so called Stern layer) while the rest extend into bulk solution to a distance determined by the double layer extension. The extension of the double layer depends on electrolyte concentration. When two droplets with their extended double layers (as is the case at low electrolyte concentration) approach a distance of separation h such that the double layers begin to overlap, repulsion occurs as a result of the increase in free energy of the whole system. In the simple case of two large droplets and low potential, Ψ_0, the repulsion interaction free energy G_E is given by the expression [128, 129],

$$G_E = 2\,\pi\,R\,\varepsilon_r\,\varepsilon_o\,\Psi_o^2 \ln\left[1 + \exp\left(-\kappa h_o\right)\right] \tag{89}$$

where ε_r is the permittivity of the medium, ε_o that of free space, and κ the Debye–Huckel parameter that is related to electrolyte concentration, C:

$$\kappa = \left(\frac{2\,Z^2\,e^2\,C}{\varepsilon_r\,\varepsilon_o\,kT}\right)^{1/2} \tag{90}$$

where Z is the valency of the electrolyte and e is the electronic charge.

Combination of van der Waals attraction and double layer repulsion results in the well-known theory of colloid stability due to Deryaguin, Landau, Verwey and Overbeek (DLVO theory) [128, 129]. The energy–distance curve is schematically represented in Figure 5.4. It is characterized by two minima and one maximum. At long distances of separation, attraction prevails resulting in a shallow minimum (secondary minimum) whose depth depends on particle size, Hamaker constant, and electrolyte concentration. The attraction energy in this minimum is usually small of the order of few kT units. In contrast, at very short distances of separation the attractive force becomes much larger than the repulsion force resulting in a deep primary minimum. If the droplets are able to reach such separation distances (i.e., in the absence of a sufficient energy barrier), very strong attraction occurs and the droplets form large aggregate units with small separation between the surfaces. This strong attraction, sometimes referred to as coagulation, is prevented by the presence of an energy maximum at intermediate distances of separation. The height of this maximum is directly proportional to the surface potential Ψ_o and inversely proportional to the electrolyte concentration. Thus, by controlling Ψ_o and C, one can make this height sufficient (> 25 kT) to prevent coagulation in the primary minimum. However, in some situations one may need to create weak attraction in the secondary minimum in order to reduce creaming or sedimentation. This is achieved by using intermediate electrolyte concentrations and using larger emulsion droplets.

The second mechanism by which flocculation may be prevented is that of steric stabilization. This is produced by using nonionic surfactants or polymers that adsorb at the liquid/liquid interface with their hydrophobic portion, leaving a thick layer of hydrophilic chains in bulk solution, e.g., polyethylene oxide (PEO) or polyvinyl alcohol (PVA). These thick hydrophilic chains produce repulsion as a result of two main effects. The first, usually referred to as mixing interaction (osmotic repulsion), results from the unfavorable mixing of the hydrophilic layers on close approach of the droplets. When the latter approach a distance of separation h that is smaller than twice the adsorbed layer thickness (2δ), overlap of these chains may occur [130]. However, when these chains are in good solvent condition (such as PEO or PVA in water) such overlap becomes unfavorable as a result of the increase of the osmotic pressure in the overlap region. This results in diffusion of solvent molecules into this overlap region, thus separating the droplets, i.e., resulting in repulsion. The free energy of repulsion due to this overlap effect can be calculated from the free energy of mixing of the two polymer layers. This results in the following expression for G_{mix} [130],

$$\frac{G_{mix}}{kT} = \frac{4\pi}{3V_1} \varphi_2^2 N_{\mathrm{av}} \left(\frac{1}{2} - \chi\right)\left(3R + 2\delta + \frac{h}{2}\right)\left(\delta - \frac{h}{2}\right)^2 \tag{91}$$

where V_1 is the molar volume of the solvent, φ_2 is the volume fraction of the polymer or surfactant in the adsorbed layer, N_{av} is Avogadro's constant, and χ is the Flory–Huggins chain-solvent interaction parameter.

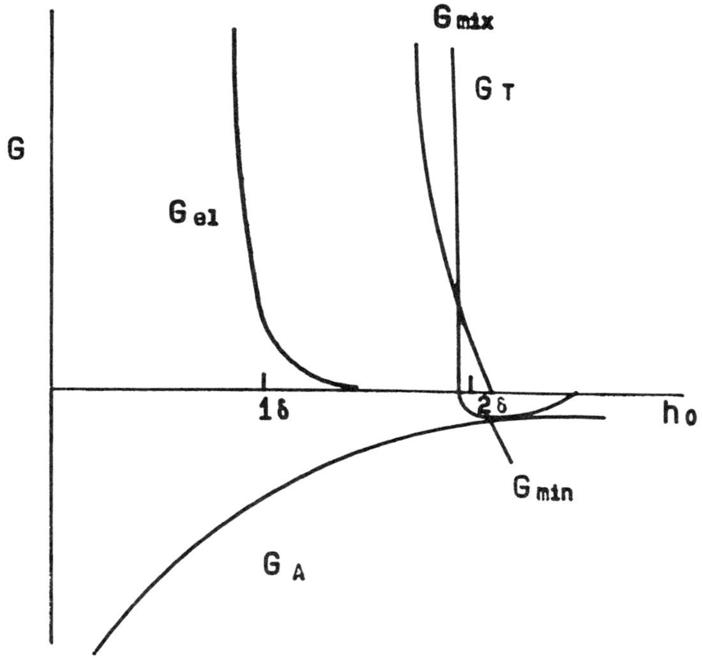

Figure 5.5 Variation of G_{mix}, G_{el}, and G_T with h (schematic).

It is clear from Eq. (91) that G_{mix} is positive (i.e., repulsive) when $\chi < 0.5$, i.e., when the chains are in good solvent conditions; when $\chi > 0.5$, G_{mix} becomes negative, i.e., attractive. There is one point at which $\chi = 0.5$ and this is referred to as the θ point for the chain that determines the onset of attraction.

The second effect that results from the presence of adsorbed layer is the loss in configurational entropy of the chains when significant overlap occurs. This effect, which is always repulsive, is usually referred to as the entropic, elastic, or volume restriction effect, G_{el}.

Combination of steric interaction with the van der Waals attraction results in an energy–distance curve as schematically represented in Figure 5.5. It can be seen that G_{mix} starts to increase rapidly as soon as h becomes smaller than 2δ. On the other hand, G_{el} begins to increase with decrease of h when the latter becomes significantly smaller than 2δ. When G_{mix} and G_{el} are combined with G_A, the total energy G_T distance curve only shows one minimum, whose location depends on 2δ and whose magnitude depends on the Hamaker constant and droplet radius R. It is clear that if 2δ is made sufficiently large and R sufficiently small, the depth of the minimum can become very small and one may approach thermodynamic stability. This explains why nonionic surfactant and polymers are relatively more effective in stabilizing emulsions against flocculation. However, one should ensure that the medium for the chains remains a good solvent, otherwise incipient flocculation occurs.

C. Ostwald Ripening

Ostwald ripening results from the finite solubility of the liquid phases. With emulsions that are polydisperse (this is usually the case) the smaller droplets will have a larger chemical potential (larger solubility) than the larger droplets. The higher solubility of the smaller droplets is the result of their higher radii of curvature (note that $S \propto 2\,\gamma/r$). With time, the smaller droplets disappear by dissolution and diffusion and become deposited on the larger droplets. This process, usually referred to as Ostwald ripening, is determined by the difference in solubility between small and large droplets as given by the Ostwald equation,

$$\frac{RT}{M}\ln\frac{S_1}{S_2} = \frac{2\gamma}{\rho}\left(\frac{1}{R_1} - \frac{1}{R_2}\right) \tag{92}$$

where S_1 is the solubility of a droplet of radius R_1 and S_2 that of a droplet with radius R_2 (note that $S_1 > S_2$ when $R_1 < R_2$), M is the molecular weight, and ρ is the density of the droplets.

The above process of Ostwald ripening is reduced by the presence of surfactants, which play two main roles. First, by adsorption of surfactants, γ is reduced, thus reducing the driving force for Ostwald ripening. Second, surfactants produce a surface tension gradient (Gibbs elasticity) that will also reduce Ostwald ripening. This can be understood from the following argument [131]. A droplet is in mechanical equilibrium if

$$\frac{dp}{dR} > R,$$

i.e., when

$$\frac{d\gamma}{d \ln R} > \gamma$$

Since

$$A = 4\pi R^2$$

then

$$\frac{2d\gamma}{d \ln A} > \gamma$$

or

$$2\varepsilon > \gamma$$

where ε is the interfacial dilational modulus. Thus, when twice the interfacial elasticity exceeds the interfaced tension, Ostwald ripening is significantly reduced.

Another method of reducing Ostwald ripening, introduced by Davies and Smith [132], is to incorporate a small proportion of a highly insoluble oil within the emulsion droplets. This reduces the molecular diffusion of the oil molecules, which are assumed to be the driving force for Ostwald ripening.

D. Emulsion Coalescence

When two emulsion droplets come in close contact in a floc or during Brownian collision, e.g., in a creamed or sedimented layer, thinning and disruption of the liquid film may occur, resulting in its eventual rupture and hence the joining together of droplets, i.e., their coalescence. The process of thinning and disruption of liquid lamellae between emulsion droplets is complex. For example, during a Brownian encounter, or in a cream or sediment, emulsion droplets may produce surface or film thickness fluctuations in the region of closest approach. The surface fluctuations produce waves that may grow in amplitude and during close approach the apexes of these fluctuations may join causing coalescence (region of high van der Waals attraction). Alternatively, any film thickness fluctuations may result in regions of small thicknesses for van der Waals attraction to cause even more thinning with the ultimate disruption of the whole film. Unfortunately, the process of coalescence is far from being well understood, although some guidelines may be obtained by considering the balance of surface forces in the liquid lamellae between the droplets. A useful picture was introduced by Deryaguin and Obucher [133] who introduced the concept of the disjoining pressure $\pi(h)$ for thin films adhering to substrates. $\pi(h)$ balances the excess normal pressure $P(h) - P_o$ in the film. $P(h)$ is the normal pressure of a film of thickness h, whereas P_o is the normal pressure of a sufficiently thick film such that the interaction free energy is zero. It should be noted that $\pi(h)$ is the net force per unit area acting across the film, i.e., normal to the interfaces. Thus $\pi(h)$ is simply equal to $-d\,V_T/dh$, where V_T is the net force that results from three main contributions, van der Waals, electrostatic, and steric forces, i.e.,

$$\pi(h) = \pi_A + \pi_E + \pi_S \tag{93}$$

For producing a stable film $\pi(h)$ needs to be positive, i.e.,

$$\pi_e + \pi_S > \pi_A$$

Thus, to reduce coalescence one needs to enhance the repulsion between the surfactant layers, e.g., by using a charged film or using surfactants with long hydrophilic chains that produce a strong steric repulsion.

For reduction of coalescence, one needs to dampen the fluctuation in surface waves or film thickness. This is produced by enhancement of the Gibbs–Marangoni effect. Several methods can be applied to reduce or eliminate coalescence and these are summarized below.

One of the earliest methods for reduction of coalescence is the use of mixed surfactant films. These have the effect of increasing the Gibbs elasticity and/or interfacial viscosity. Both effects result in reduction of film fluctuations and hence reduction of coalescence. In addition, mixed surfactant films are usually more condensed and hence diffusion of the surfactant molecules from the interface is greatly hindered. An alternative explanation for enhanced stability using surfactant mixture was introduced by Friberg and coworkers [134] who considered the formation of a three-dimensional association structure (liquid crystals) at the oil/water interface. The presence of these liquid crystalline structures prevents coalescence since one has to remove several surfactant layers before droplet–droplet contact may occur.

Another method of reducing coalescence is the use of macromolecular surfactants such as gums, proteins, and synthetic polymers, e.g., A-B or A-B-A block copolymer. The latter in particular could produce very stable films by strong adsorption of the B groups of the molecules leaving the A chains dangling in solution and providing a strong steric barrier that prevents any coalescence. Examples of such molecules are PVA and, PEO and polypropylene oxide block copolymers.

E. Phase Inversion

Phase inversion is the process whereby the internal and external phases of an emulsion suddenly invert, i.e., o/w to w/o and vice versa. Phase inversion can be easily observed if the oil volume fraction of, say, an o/w emulsion is gradually increased. For example, at a given emulsifier concentration, it is often observed that the viscosity of an emul-

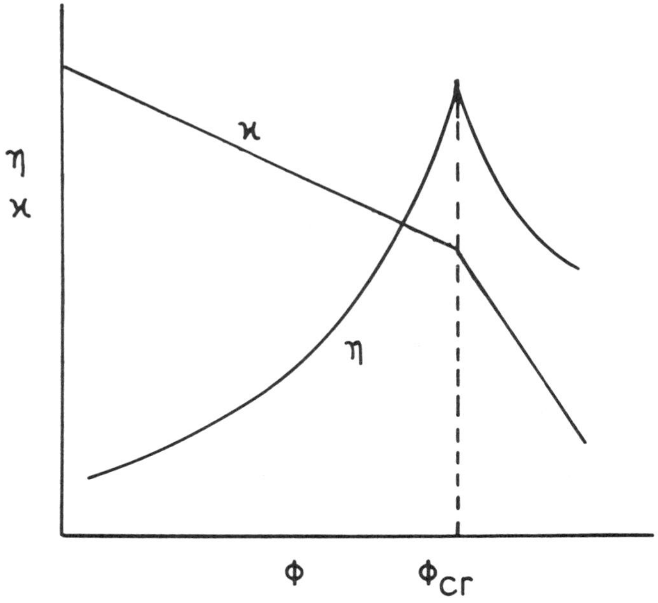

Figure 5.6 Schematic representation of variation of η and κ with φ for an o/w emulsion.

sion gradually increases with increase in φ, but at a certain critical volume fraction, φ_{cr}, there is a sudden decrease. The same sudden change is observed in the specific conductivity κ, which initially decreases slowly with increase in φ, but decreases much more rapidly above φ_{cr}. This is illustrated in Figure 5.6 which schematically shows the variation of η and κ with φ. The critical volume fraction corresponds to the point at which the o/w emulsion inverts to a w/o emulsion. The sharp decrease in η observed at the inversion point is due to the sudden reduction in disperse phase volume fraction. The sudden rapid decrease in κ is due to the fact that the emulsion now becomes oil continuous. This has a much lower conductivity than the aqueous continuous phase emulsion.

In the early theories of phase inversion, it was postulated that the inversion takes place as a result of difficulty in packing emulsion droplets above a certain volume fraction (the maximum packing fraction). For example, if the emulsion is monodisperse $\varphi_p = 0.74$ and any attempt to increase φ above this value leads to inversion. However, a number of investigations have clearly indicated the invalidity of this argument, with inversion being found to occur at values much greater or smaller than 0.74. At present, there does not seem to be any quantitative theory that explains phase inversion. However, location of the inversion point is of practical importance particularly on storage of the emulsion. As mentioned above, the PIT can be an important criterion for assessment of the long term physical stability of emulsions.

V. CHARACTERIZATION OF EMULSIONS AND ASSESSMENT OF THEIR LONG-TERM PHYSICAL STABILITY

For the characterization of emulsion systems, it is necessary to obtain fundamental information on the liquid/liquid interface (e.g., interfacial tension and interfacial rheology) and properties of the bulk emulsion system such as droplet size distribution, flocculation, coalescence, phase inversion, and rheology. The information obtained, if analyzed carefully, can be used for the assessment and (in some cases) for the prediction of the long term physical stability of the emulsion.

A. Interfacial Properties

As discussed before, measurement of the interfacial tension is a fundamental property that gives information on surfactant and/or polymer adsorption and its kinetics. The results obtained can give valuable information on the orientation and conformation of the molecules at the interface (see Chapter 3). Moreover, measurements under dynamic conditions (i.e., on expansion of the interface) allows one to obtain the Gibbs elasticity, which may be a crucial factor for the coalescence process. Below a brief description of the various techniques that may be used for measurement of the interfacial tension is given. This is followed by a short section on interfacial rheology.

1. Interfacial Tension Measurements

These methods may be classified into two categories: those in which the properties of the meniscus is measured at equilibrium, e.g., pendent drop or sessile drop profile and Wilhelmy plate methods, and those where the measurement is made under nonequilibrium or quasi-equilibrium conditions such as the drop volume (weight) or the du Nouy ring method. The latter methods are faster although they suffer from the disadvantage of premature rupture and expansion of the interface, causing adsorption depletion. They are also unsuitable for measurement of the interfacial tension in the presence of macromolecules, since in this case equilibrium may require hours or even days. For measurement of low interfacial tensions (< 0.1 mN m^{-1}) the spinning drop technique is applied. Below a brief description of each of these techniques is given.

(a) The Wilhelmy Plate Method

In this method [135] a thin plate made from glass (e.g., a microscope cover slide) or platinum foil is either detached from the interface (nonequilibrium condition) or its weight measured statically using an accurate microbalance. In the detachment method, the total force F is given by the weight of the plate W and the interfacial tension force,

$$F = W + \gamma p \tag{94}$$

where p is the "contact length" of the plate with the liquid, i.e., the plate perimeter. Provided the contact angle of the liquid is zero, no correction is required for Eq. (94). Thus, the Wilhelmy plate method can be applied in the same manner as the du Nouy technique described below.

The static technique may be applied for following the interfacial tension as a function of time (to follow the kinetics of adsorption) until equilibrium is reached. In this case, the plate is suspended from one arm of a microbalance and allowed to penetrate the upper liquid layer (usually the oil) until it touches the interface, or alternatively the whole vessel containing the two liquid layers is raised until the interface touches the plate. The increase in weight ΔW is given by as follows:

$$\Delta W = \gamma p \cos \theta \tag{95}$$

where θ is the contact angle. If the plate is completely wetted by the lower liquid as it penetrates, $\theta = 0$ and γ may be calculated directly from ΔW. Care should always be taken that the plate is completely wetted by the aqueous solution. For that purpose, a roughened platinum or glass plate is used to ensure a zero contact angle. However, if the oil is denser than water, a hydrophobic plate is used so that when the plate penetrates through the upper aqueous layer and touches the interface it is completely wetted by the oil phase.

(b) The Pendent and Drop Method

If a drop of oil is allowed to hang from the end of a capillary that is immersed in the aqueous phase, it will adopt an equilibrium profile shown in Figure 5.7 that is a unique function of the tube radius, the

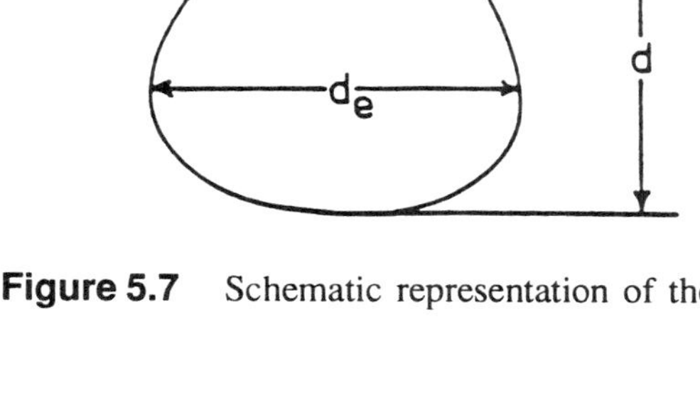

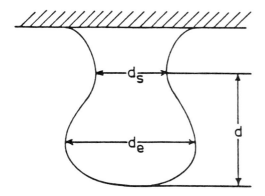

Figure 5.7 Schematic representation of the profile of a pendent drop.

interfacial tension, the density, and the gravitational field. The interfacial tension is given by the following [136]:

$$\gamma = \frac{\Delta\rho\, g\, d_e^2}{H} \tag{96}$$

where $\Delta\rho$ is the density difference between the two phases, d_e is the equatorial diameter of the drop (see Fig. 5.7), and H is a function of d_s/d_e, where d_s is the diameter measured at a distance d from the bottom of the drop (see Fig. 5.7). The relationship between H and the experimental values of d_s/d_e has been obtained empirically using pendent drops of water. Accurate values of H have been obtained by Niederhauser and Bartell [137].

(c) The Du Nouy Ring Method

Basically one measures the force required to detach a ring or loop of wire from the liquid/liquid interface [138]. As a first approximation, the detachment force is taken to be equal to the interfacial tension γ multiplied by the perimeter of the ring, i.e.,

$$F = W + 4\pi R \gamma \tag{97}$$

where W is the weight of the ring. Harkins and Jordan [139] introduced a correction factor f (that is a function of meniscus volume V and radius r of the wire) for more accurate calculation of γ from F, i.e.,

$$\begin{aligned} f &= \frac{\gamma}{\gamma_{ideal}} \\ &= f\left(\frac{R^3}{V}, \frac{R}{r}\right) \end{aligned} \tag{98}$$

Values of the correction factor f were tabulated by Harkins and Jordan [139] and by Fox and Chrisman [140]. Some theoretical account of f was given by Freud and Freud [140].

When using the du Nouy method for measurement of γ one must be sure that the ring is kept horizontal during measurement. Moreover, the ring should be free from any contaminant and this is usually achieved by using a platinum ring that is flamed before use.

(d) The Drop Volume (Weight) Method

Here one determines the volume V (or weight W) of a drop of liquid (immersed in the second less dense liquid) that becomes detached from a vertically mounted capillary tip having a circular cross-section of radius r. The ideal drop weight W_{ideal} is given by the expression,

$$W_{ideal} = 2\,\pi\,r\,\gamma \tag{99}$$

In practice, a weight W is obtained which is less than W_{ideal} because a portion of the drop remains attached to the tube tip. Thus, Eq. (99) should include a correction factor φ, that is a function of the tube radius r and some linear dimension of the drop, i.e., $V^{1/3}$. Thus,

$$W = 2\,\pi\,r\,\gamma\,\varphi\left(\frac{r}{V^{1/3}}\right) \tag{100}$$

values of $(r/V^{1/3})$ have been tabulated by Harkins and Brown [141]. Lando and Oakley [142] used a quadratic equation to fit the correction function to $(r/V^{1/3})$. A better fit was provided by Wilkinson and Kidwell [143].

(e) The Spinning Drop Method

The spinning drop method is particularly useful for the measurement of very low interfacial tensions ($< 10^{-1}$ mN m^{-1}), which are particularly important in applications such as spontaneous emulsification (see Chapter 4) and the formation of microemulsions (see Chapter 7). Such low interfacial tensions may also be reached with emulsions particularly when mixed surfactant films are used. A drop of the less dense liquid A is suspended in a tube containing the second liquid B. On rotating the whole mass (Figure 5.8) the drop of the liquid moves to the center. With increasing speed of revolution, the drop elongates as the centrifugal force opposes the interfacial tension force that tends to maintain the spherical shape, i.e., that having minimum surface

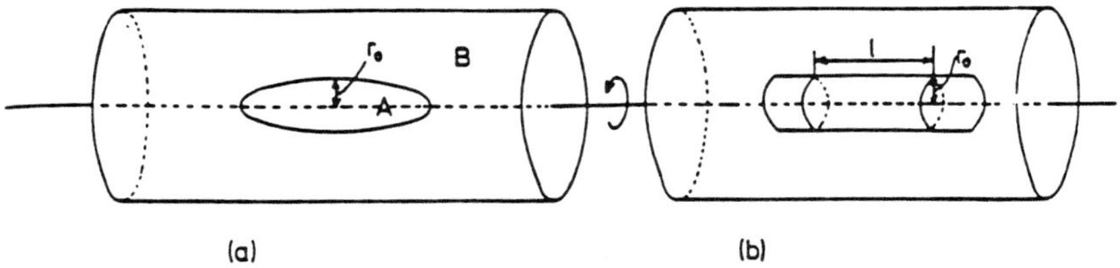

Figure 5.8 Schematic representation of a spinning drop. (a) Prolate ellipsoid, (b) elongated cylinder.

area. An equilibrium shape is reached at any given speed of rotation. At moderate speeds of rotation the drop approximates to a prolate ellipsoid, whereas at very high speeds of revolution the drop approximates to an elongated cylinder. This is schematically shown in Figure 5.8.

When the shape of the drop approximates a cylinder (Fig. 5.8b), the interfacial tension is given by the following expression [144]:

$$\gamma = \frac{\omega^2 \, \Delta\rho \, r_o^4}{4} \tag{101}$$

where ω is the speed of rotation, $\Delta\rho$ the density difference between the two liquids A and B, and r_o the radius of the elongated cylinder. Equation (101) is valid when the length of the elongated cylinder is much more than r_o.

B. Interfacial Rheology

As mentioned above, interfacial elasticity and viscosity may play an important role in prevention of coalescence and Ostwald ripening of emulsions. These parameters may be measured at a planar oil/water interface using specialized techniques. A brief description of how one can measure the interfacial elasticity and interfacial viscosity of adsorbed films is given below.

The surface viscosity of an interfacial film is the ratio between the shear stress and shear rate in the plane of the interface, i.e., it is a two-dimensional viscosity. The unit for surface viscosity is therefore Nm s^{-1} or surface Pas. Basically the interfacial viscosity is the product of bulk viscosity by the thickness of the interfacial region. A liquid–liquid interface has viscosity if the interface itself contributes to the resistance of shear in the plane of the interface [145]. Pure liquids in equilibrium with air or their vapor do not show much viscosity effect. However, most surfactants and their mixtures and macromolecules adsorbed at the interface are viscous, showing a high surface-induced viscosity. Usually the surface viscosity is much higher (by orders of magnitude) than the bulk viscosity. The high viscosity of adsorbed films can be easily accounted for in terms of orientation of the molecules at the interface. For example, surfactants orient at the o/w interface with the hydrophobic portions pointing to (or dissolved in) the oil phase and the polar groups pointing to the aqueous phase. Such films resist compression by a film pressure, π, given by the expression:

$$\pi = \gamma_o - \gamma \tag{102}$$

where γ_o is the interfacial tension of the clean surface adsorption, i.e., before adsorption of a surfactant or polymer (of the order of 30–50 mN m^{-1}) and γ is the interfacial tension after adsorption that can reach values of a fraction of 1 mN m^{-1}, Thus surface pressures of the order of 30–50 mN m^{-1} can be obtained that would correspond to a high surface viscosity. As the surfactant film is compressed at the interface, the surfactant molecules become more closely packed and oriented more nearly normal to the surface, thus producing a higher surface viscosity.

Macromolecular films also give a high interfacial viscosity due to orientation of the chains at the liquid/liquid interface. Usually the macromolecule adopts a train-loop-tail configuration (see Chapter 3) and the macromolecule resists compression as a result of lateral repulsion between the loops and/or tails.

The simplest procedure to measure a surface viscosity is to use a torsion pendulum viscometer. In this case, the damping of a torsion

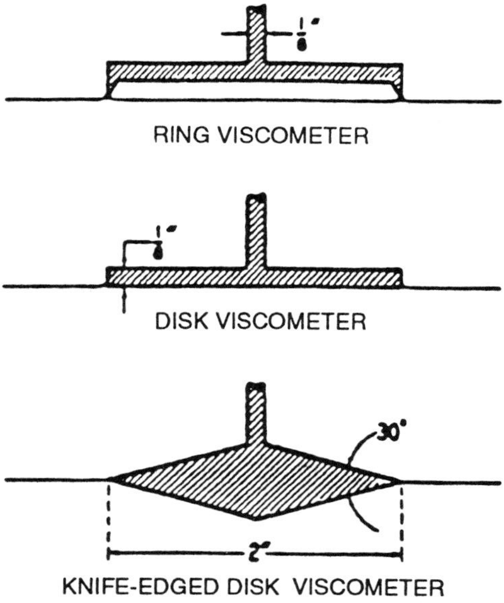

Figure 5.9 Surface viscometer designs.

pendulum due to the viscous drag of a surface film is measured. The shearing element can be in the form of a ring, a disk, or knife-edged disk, which is suspended by a torsion element and positioned at the place of the interface. This is illustrated in Figure 5.9.

Measurements are made of the period of the pendulum and the damping as the pendulum oscillates. The apparent surface viscosity, η_s, is given by

$$\eta_s = \eta_o \left(\frac{\Delta/\Delta_o}{t/t_o} - 1 \right) \tag{103}$$

where η_o is the sum of the bulk viscosities of the two phases forming the interface, Δ is the difference in logarithm of the amplitudes of successive swings for the interface with adsorbed surfactant or polymer, Δ_o is the corresponding value without surfactant or polymer, t is the period of the pendulum for the film-covered interface, and t_s is the corresponding value without the adsorbed film. The surface viscosity, η_s, is expressed in surface Pas by the expression [145],

$$\eta_s = \frac{C_w I}{2\pi} \frac{R_2^2 - R_1^2}{R_1^2 R_2^2} \left(\frac{\Delta}{7.4 + \Delta^2} - \frac{\Delta_o}{7.4 + \Delta_o} \right) \tag{104}$$

where C_w is the torsional modulus of the wire, I is the polar moment of inertia of the oscillating pendulum, R_1 is the radius of the surface viscometer, and R_2 is the radius of the container.

One of the main disadvantages of the torsion pendulum viscometer is that it uses a range of shear rates and amplitudes in each determination. Thus, the surface viscosity must be considered as an average value over a range of shear rates. In spite of this disadvantage, the pendulum surface viscometer enables one to detect and measure the surface viscosity of Newtonian films and to explore the stability and shear rate dependence of non-Newtonian films.

For more accurate shear rate dependence of surface viscosity, the rotational torsion surface viscometer is perhaps the most convenient. The surface film is sheared between rotating rings on the interface. The surface shear rate can be held constant by rotating one ring at any desired rate and measuring the torque on the other ring. In this case the surface viscosity is given by [145],

$$\eta_s = \frac{SKt}{8\pi^2} \frac{R_2^2 - R_1^2}{R_1^2 R_2^2} \tag{105}$$

where t is the time of revolution of the ring, K is the torsional movement corresponding to 1° strain, and S is the difference in degrees strain for the same velocity gradient in the presence and absence of a surface film. An alternative equation may be used for η_s in terms of the torque T and the angular velocity Ω in the same manner as for rotational viscometers, i.e.,

$$\eta_s = \frac{T}{4\pi\Omega} \frac{R_1^2 - R_2^2}{R_1^2 R_2^2} + \frac{\sigma_s}{\Omega} \ln \frac{R_2}{R_1} \tag{106}$$

where σ_s is the surface yield stress.

Although rotational surface viscometers provide great accuracy since the shear rate is well defined, they are not sensitive enough for measurement of films with low surface viscosities, such as those encountered with surfactants. They are more suitable for measurement of the surface viscosity of polymeric films which have higher surface viscosities than surfactant film. They are also very suitable for investigation of non-Newtonian films.

As mentioned above, interfacial films show both viscosity and elasticity. Films are elastic if they resist deformation in the plane of the interface and if the surface tends to recover its natural shape when the deforming force is removed. Similar to bulk materials, interfacial elasticity can be measured using static and dynamic methods (see below). Furthermore, the elastic constant of the surface film depends on the nature of the deforming stress. If the area of the film is held constant and static measurements are made of resistance to deformation in the plane of the interface, the surface elasticity, E_s, is given by

$$E_s = \frac{C_w}{4\pi} \frac{\Omega_w}{\Omega_f} \left(\frac{1}{R_1^2} - \frac{1}{R_2^2} \right) \tag{107}$$

where Ω_f and Ω_w are the angular displacements in radians of the film and wire respectively.

In dynamic measurements, the surface shear modulus, G_s, is given by

$$G_s = \pi I \left(\frac{1}{t^2} - \frac{1}{t_o^2} \right) \left(\frac{1}{R_1^2} - \frac{1}{R_2^2} \right) \tag{108}$$

where I is the moment of inertia and t and t_o are the periods of the pendulum at the interface in the presence and absence of a surface film, respectively.

The dilational elasticity, ε, is given by,

$$\varepsilon = \frac{d\gamma}{d \ln A} \tag{109}$$

where A is the area of the interface and γ is the interfacial tension. The interfacial elasticity can be measured using a Langmuir trough with two movable barriers. By moving the two barriers simultaneously, the area of the interface is changed (a relatively small increase in the area, less than 1%, is usually performed) and the interfacial tension is measured using a Wilhelmy plate placed at the o/w interface. The plate is placed in the middle of the trough far from the movable barriers.

The dilational elasticity can be also measured by application of surface waves, with frequency ω. The dilational elasticity is given by the expression

$$\varepsilon = \frac{\varepsilon_o\,[1 + (\tau/\omega_o)^{1/2}]}{[1 + 2(\tau/\omega_o)^{1/2} + 2(\tau/\omega_o)]} \tag{110}$$

where ω_o is the Gibbs elasticity and τ a "diffusion parameter" related to the diffusion coefficient of the surfactant molecule.

6

Suspension Concentrates

I. INTRODUCTION

The formulation of agrochemicals as dispersions of solids in aqueous solution (to be referred to as suspension concentrates or SCs) has attracted considerable attention in recent years. Such formulations are natural replacement to wettable powders (WPs). The latter are produced by mixing the active ingredient with a filler (usually a clay material) and a surfactant (dispersing and wetting agent). These powders are dispersed into the spray tank to produce a coarse suspension that is applied to the crop. Although WPs are simple to formulate they are not the most convenient for the farmer. Apart from being dusty (and occupying a large volume due to their low bulk density), they tend to settle fast in the spray tank. Moreover, they do not provide optimum biological efficiency as a result of the large particle size of the system. In addition, one cannot incorporate the necessary adjuvants (mostly surfactants) in the formulation. These problems are overcome by formulating the agrochemical as an aqueous SC.

Several advantages may be quoted for SCs. First, one may control the particle size by controlling the milling conditions and making the proper choice of dispersing agent. Second, it is possible to incorporate high concentrations of surfactants in the formulation, which is sometimes essential for enhancing wetting, spreading, and penetration (see Chapter 8). Stickers may also be added to enhance adhesion and in some cases to provide slow release.

In recent years there has been considerable research on the factors that govern the stability of suspension concentrates [147–149]. The theories of colloid stability could be applied to predict the physical states of these systems on storage. In addition, analysis of the problem of sedimentation of SCs at a fundamental level has been undertaken [150]. Since the density of the particles is usually higher than that of the medium (water), SCs tend to separate as a result of sedimentation. The sedimented particles tend to form a compact layer at the bottom of the container (sometimes referred to as clay or cake), which is very difficult to redisperse. It is therefore essential to reduce sedimentation and formation of clays by incorporation of an antisettling agent.

In this chapter, I will attempt to address the above-mentioned phenomena at a fundamental level. The chapter will start with a section on the preparation of suspension concentrates and the role of surfactants (dispersing agents). This is followed by a section on the control of the physical stability of suspensions. The problem of Ostwald ripening (crystal growth) will also be briefly described and particular attention will be paid to the role of surfactants. The next section will deal with the problem of sedimentation and prevention of claying. The various methods that may be applied to reduce sedimentation and prevention of the formation of hard clays will be summarized. The last section will deal with the methods that may be applied for the assessment of the physical stability of SCs. For the assessment of flocculation and crystal growth, particle size analysis techniques are commonly applied. The bulk properties of the suspension, such as sedimentation and separation, as well as redispersion on dilution, may be assessed using rheological techniques. The latter will be summarized with particular emphasis on their application in prediction of the long-term physical stability of suspension concentrates.

II. PREPARATION OF SUSPENSION CONCENTRATES AND THE ROLE OF SURFACTANTS

Suspension concentrates are usually formulated using a wet-milling process that requires the addition of a surfactant (dispersing agent). The latter should satisfy the following criteria: be a good wetting agent for the agrochemical powder (both external and internal surfaces of the powder aggregates or agglomerates must be spontaneously wetted); be a good dispersing agent to break such aggregates or agglomerates into smaller units and subsequently help in the milling process (one usually aims for a dispersion with a volume mean diameter of 1–2 μm); and provide good stability in the colloid sense (this is essential for maintaining the particles as individual units once formed). Powerful dispersing agents are particularly important for the preparation of highly concentrated suspensions sometimes required for seed dressing. Any flocculation will cause a rapid increase in the viscosity of the suspension and this makes the wet milling of the agrochemical a difficult process. This section will discuss the wetting of agrochemical powders, their subsequent dispersions, and milling. The subject of colloid stability will be dealt with in the next section.

A. Wetting of the Agrochemical Powder

Dry powders of organic compounds usually consist of particles of various degrees of complexity, depending on the isolation stages and the drying process. Generally, the particles in a dry powder form aggregates (in which the particles are joined together with their crystal faces) or agglomerates (in which the particles touch at edges or corners) forming a looser more open structure. It is essential in the dispersion process to wet the external as well as the internal surfaces and displace the air entrapped between the particles. This is usually achieved by the use of surface active agents of the ionic or nonionic type. In some cases, macromolecules or polyelectrolytes may be efficient in this wetting process. This may be the case since these polymers contain a very wide distribution of molecular weights and the low molecular weight fractions may act as efficient wetting agents.

For efficient wetting the molecules should lower the surface tension of water (see below) and they should diffuse fast in solution and become quickly adsorbed at the solid/solution interface.

Wetting of a solid is usually described in terms of the equilibrium contact angle θ and the appropriate interfacial tensions, using the classical Young's equation:

$$\gamma_{sv} - \gamma_{SL} = \gamma_{LV} \cos \theta \quad (111)$$

or

$$\cos \theta = \frac{\gamma_{SV} - \gamma_{SL}}{\gamma_{LV}} \quad (112)$$

where γ represents the interfacial tension and the symbols S, L, and V refer to the solid, liquid, and vapor, respectively. It is clear from Eq. (111) that if $\theta < 90°$, a reduction in γ_{LV} improves wetting. Hence the use of surfactants that reduce both γ_{LV} and γ_{SL} to aid wetting is clear. However, the process of wetting of particulate solids is more complex and involves at least three distinct types of wetting [151, 152], i.e., adhesional wetting, spreading wetting, and immersional wetting. This is illustrated in Figure 6.1, which describes schematically the wetting of a cube of a solid by a liquid. Assuming that the surface area of the cube is unity, then the work of dispersion is given by,

$$\begin{aligned} W_d &= W_a + W_i + W_s \\ &= -6\gamma_{SL} - 6\gamma_{SV} \\ &= -6\gamma_{LV} \cos \theta \end{aligned} \quad (113)$$

Thus, the wetting of a solid by a liquid depends on two measurable quantities, γ_{LV} and θ, and hence Eq. (112) may be used to predict whether the process is spontaneous, i.e., W_d is negative. The adhesion process is invariably spontaneous, whereas the other two processes

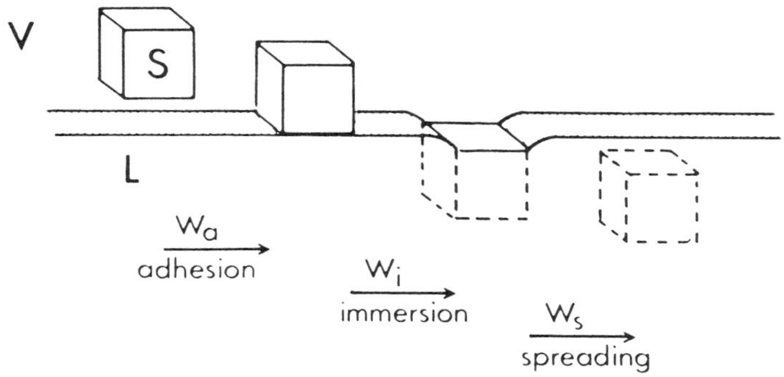

Figure 6.1 The three stages involved in the complete wetting of a solid cube by a liquid.

depend on the value of θ. For example, spreading is only spontaneous when θ = 0, whereas immersion and dispersion are spontaneous when θ < 90°.

The next stage to be considered is the wetting of the internal surface, which implies penetration of the liquid into channels between and inside the agglomerates. This is more difficult to define precisely. However, one may make use of the equation derived for capillary phenomena. To force a liquid into a capillary tube of radius r, pressure P is required such that

$$P = -\frac{2\gamma_{LV}\cos\theta}{r} = -\frac{2(\gamma_{SV} - \gamma_{SL})}{r} \tag{114}$$

Equation (114) shows that to increase penetration, θ and γ_{SL} have to be made as small as possible, e.g., by sufficient adsorption of surfactant. However, when θ = 0, P is proportional to γ_{LV}, i.e., a large surface tension is required. These two opposing effects show that the proper choice of a surfactant is not simple. In most cases a compromise has to be made to minimize θ while not having a too small surface tension to aid penetration.

Another important factor in the wetting process is the role of penetration of the liquid into the channels between and inside the agglomerates. This has to be as fast as possible to aid the dispersion process. Penetration of liquids into powders can be qualitatively treated by the Rideal–Washburn equation [153], assuming the powder to be represented by a bundle of capillaries. For horizontal capillaries (where gravity may be neglected), the rate of penetration of a liquid into an air filled tube is given by the equation:

$$\frac{dl}{dt} = \frac{r\gamma\cos\theta}{4\eta l} \tag{115}$$

where l is the distance the liquid has traveled along the pore in time t, r is the radius of the capillary tube, and γ and η are the surface tension and viscosity of the liquid, respectively. Thus, to enhance penetration, γ_{LV} should be made as high as possible and θ as low as possible. Moreover, to increase penetration, the powder has to be made as loose as possible.

Integration of Eq. (115) leads to

$$l^2 = \frac{r\gamma_{LV}\cos\theta\, t}{2\eta} \tag{116}$$

which has to be modified for powders by replacing r with a factor K that contains an "effective radius" for the bed and a tortuosity factor to allow for the random shape and size of the capillaries, i.e.,

$$l^2 = \frac{K\gamma_{LV}\cos\theta\, t}{2\eta} \tag{117}$$

Equation (117) forms the basis of the method commonly used for measuring θ for a powder/liquid system. A known weight of the dry powder is packed in a glass tube fitted at one end with a sintered glass disk and the rate of rise of the liquid into the powder bed is measured. A plot of l^2 vs. t is usually linear with a slope equal to

$$\frac{K \gamma_{LV} \cos \theta}{2\eta}$$

The value of K may be obtained by using a liquid of known surface tension that gives zero θ.

Thus, in summary, the dispersion of a powder in a liquid depends on three main factors: the energy of wetting of the external surface, the pressure involved in the liquid penetrating inside and between the agglomerates, and the rate of penetration of the liquid into the powder. All these factors are related to two main parameters, namely, γ_{LV} and θ. In general, the process is likely to be more spontaneous the lower the θ and the higher γ_{LV}. Since these two factors tend to operate in opposite senses, choosing of the proper surfactant (dispersing agent) can be a difficult task.

B. Dispersion and Milling

For the dispersion of aggregates and agglomerates into smaller units one requires high-speed mixing, e.g., a Silverson mixer. In some cases the dispersion process is easy and the capillary pressure may be sufficient to break up the aggregates and agglomerates into primary units. The process is aided by the surfactant, which becomes adsorbed on the particle surface. However, one should be careful during the mixing process not to entrap air (foam), which causes an increase in the viscosity of the suspension and prevents easy dispersion and subsequent grinding. If foam formation becomes a problem, one should add antifoaming agents such as polysiloxane surfactants.

After completion of the dispersion process, the suspension is transferred to a ball or bead mill for size reduction. Milling or comminution (the generic term for size reduction) is a complex process and there is little fundamental information on its mechanism. For the breakdown of single crystals into smaller units, mechanical energy is required. This energy in a bead mill, for example, is supplied by impaction of the glass beads with the particles. As a result, permanent deformation of the crystals and crack initiation result. This will eventually lead to the fracture of the crystals into smaller units. However, since the milling conditions are random, it is inevitable that some particles receive impacts far in excess of those required for fracture, whereas others receive impacts that are insufficient to fracture them. This makes the milling operation grossly inefficient and only a small fraction of the applied energy is actually used in comminution. The rest of the energy is dissipated as heat, vibration, sound, interparticulate friction, friction between the particles and beads, and elastic deformation of unfractured particles. For these reasons, milling conditions are usually established by a trial and error procedure. Of particular importance is the effect of various surface active agents and macromolecules on the grinding efficiency. The role played by these agents in the comminution process is far from being understood, although Rehbinder and collaborators [154–156] have given this problem particular consideration. As a result of adsorption of surfactants at the solid/liquid interface, the surface energy at the boundary is reduced and this facilitates the process of deformation or destruction. The adsorption of the surfactant at the solid/solution interface in cracks facilitates their propagation. This is usually referred to as the Rehbinder effect [157]. The surface energy manifests itself in destructive processes on solids, since the generation and growth of cracks and separation of one part of a body from another is directly connected with the development of new free surface. Thus, as a result of adsorption of surface active agents at structural defects in the surface of the crystals, fine grinding is facilitated. In the extreme case where there is a very great reduction in surface energy at the sold/liquid boundary, spontaneous dispersion may take place with the result of the formation of colloidal particles (< 1 μm). Rehbinder [158] developed a theory for such spontaneous dispersion. Unfortunately, there are not sufficient experimental data to prove or disprove the Rehbinder effect.

III. CONTROL OF THE PHYSICAL STABILITY OF SUSPENSION CONCENTRATES

A. Stability Against Aggregation

The control of stability against irreversible flocculation (where the particles are held together into aggregates that cannot be redispersed by shaking or on dilution) is achieved by the use of powerful dispersing agents, e.g., surfactants of the ionic or nonionic type, nonionic polymers or polyelectrolytes. These dispersing agents must be strongly adsorbed onto the particle surfaces and fully cover them. With ionic surfactants, irreversible flocculation is prevented by the repulsive force generated from the presence of an electrical double layer at the particle/ solution interface (see Chapter 5). Depending on the conditions, this repulsive force can be made sufficiently large to overcome the ubiquitous van der Waals attraction between the particles, at intermediate distances of separation. With nonionic surfactants and macromolecules, repulsion between the particles is ensured by the steric interaction of the adsorbed layers on the particle surfaces (see Chapter 5). With polyelectrolytes, both electrostatic and steric repulsion exist. Below a summary of the role of surfactants in stabilization of particles against flocculation is described.

Ionic surfactants such as sodium dodecyl benzenesulfonate (NaDBS) or cetyltrimethylammonium chloride (CTACl) adsorb on hydrophobic particles of agrochemicals as a result of the hydrophobic interaction between the alkyl group of the surfactant and the particle surface. As a result, the particle surface will acquire a charge that is compensated by counterions (Na^+ in the case of NaDBS and Cl^- in the case of CTACl) forming an electrical double layer.

The adsorption of ionic surfactants at the solid/solution interface is of vital importance in determining the stability of suspension concentrates. As discussed in Chapter 3, the adsorption of ionic surfactants on solid surfaces can be directly measured by equilibrating a known amount of solid (with known surface area) with surfactant solutions of various concentrations. After reaching equilibrium, the solid particles are removed (e.g., by centrifugation) and the concentration of surfactant in the supernatant liquid is analytically determined. From the difference between the initial and final surfactant concentrations (C_1 and C_2, respectively) the number of moles of surfactant adsorbed, Γ, per unit area of solid is determined and the results may be fitted to a Langmuir isotherm:

$$\Gamma = \frac{\Delta C}{mA}$$

$$= \frac{abC_2}{1 + bC_2} \tag{118}$$

where

$$\Delta C = C_1 - C_2$$

m is the mass of the solid with surface area A, a is the saturation adsorption, and b is a constant that is related to the free energy of adsorption

$$b \propto \exp - \frac{\Delta G}{RT}$$

From a, the area per surfactant ion on the surface can be calculated (area per surfactant ion = $1/a\ N_{av}$).

Results on the adsorption of ionic surfactants on pesticides are scarce. However, Tadros [159] obtained some results on the adsorption of NaDBS and CTABr on the fungicide, namely ethirimol. For NaDBS, the shape of the isotherm was of a Langmuir type, giving an area/DBS^- at saturation of about 0.14 nm^2. The adsorption of CTA^+ showed a two step isotherm with areas/CTA^+ of 0.27 and 0.14 nm^2, respectively. These results suggest full saturation of the surface with surfactant ions that are vertically oriented.

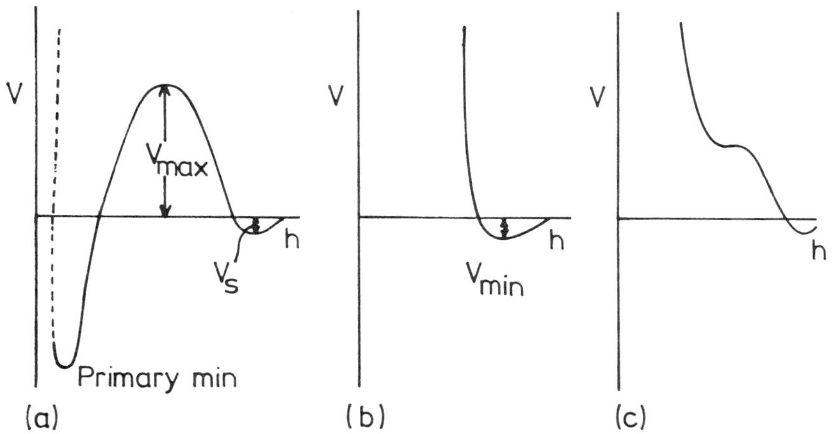

Figure 6.2 Energy–distance curves for three different stabilization mechanisms: (a) electrostatic; (b) steric; (c) electrosteric.

The above discussion shows that ionic surfactants can be used to stabilize agrochemical suspensions by producing sufficient electrostatic repulsion. When two particles with adsorbed surfactant layers approach each other to a distance where the electrical double layers begin to overlap, strong repulsion occurs preventing any particle aggregation (see Chapter 5). The energy–distance curve for such electrostatically stabilized dispersions is schematically shown in Figure 6.2a. This shows an energy maximum, that if high enough (> 25 kT) prevents particle aggregation into the primary minimum. However, ionic surfactants are the least attractive dispersing agents for the following reasons. Adsorption of ionic surfactants is seldom strong enough to prevent some desorption with the result of production of "bare" patches that may induce particle aggregation. The system is also sensitive to ionic impurities present in the water used for suspension preparation. In particular, the system will be sensitive to bivalent ions (Ca^{2+} or Mg^{2+}) that produce flocculation at relatively low concentrations.

Nonionic surfactants of the ethoxylate type, e.g., $R(CH_2CH_2O)_nOH$ or $RC_6H_5(CH_2CH_2O)_nOH$, provide a better alternative provided the molecule contains sufficient hydrophobic groups to ensure their adsorption and enough ethylene oxide units to provide an adequate energy barrier. As discussed in Chapter 5, the origin of steric repulsion arises from two main effects. The first effect arises from the unfavorable mixing of the polyethylene oxide (PEO) chains which are in good solvent condition (water as medium). This effect is referred to as the mixing or osmotic repulsion. The second effect arises from the loss in configurational entropy of the chains when these are forced to overlap on approach of the particles. This is referred to as the elastic or volume restriction effect. The energy–distance curve for such systems (Fig. 6.2b) clearly demonstrates the attraction of steric stabilization. Apart from a small attractive energy minimum (which can be reasonably shallow with sufficiently long PEO chains), strong repulsion occurs and there is no barrier to overcome. A better option is to use block and graft copolymers (polymeric surfactants) consisting of A and B units combined in A-B, A-B-A, or BA_n fashion. B represents units with high affinity for the particle surface and basically insoluble in the continuous medium, thus providing strong adsorption ("anchoring units"). A, on the other hand, represents units with high affinity to the medium (high chain solvent interaction) and little or no affinity to the particle surface. An example of such powerful dispersant is a graft copolymer of polymethyl methacrylate–methacrylic acid (the anchoring portion) and methoxypolyethylene oxide (the stabilizing chain) methacrylate [160]. Adsorption measurements of such a polymer on the pesticide ethirimol (a fungicide) showed a high affinity isotherm with no desorption. Using such a macromolecular surfactant, a suspension of high volume fractions could be prepared.

The third class of dispersing agents that is commonly used in SC formulations is that of polyelectrolytes. Of these, sulfonated naphthalene-formaldehyde condensates and lignosulfonates are the most commonly used in agrochemical formulations. These systems show a combined electrostatic and steric repulsion; the energy–distance curve is schematically illustrated in Figure 6.2c, which shows a shallow minimum and maximum at intermediate distances (characteristic of electrostatic repulsion) as well as strong repulsion at relatively short distances (characteristic of steric repulsion). The stabilization mechanism of polyelectrolytes is sometimes referred to as electrosteric. These polyelectrolytes offer some versatility in SC formulations. Since the interaction is fairly long range in nature (due to the double layer effect), one does not obtain the "hard-sphere" type of behavior that may lead to the formation of hard sediments. The steric repulsion ensures the colloid stability and prevents aggregation on storage.

B. Ostwald Ripening (Crystal Growth)

There are several ways in which crystals can grow in an aqueous suspension. One of the most familiar is the phenomenon of Ostwald ripening, which occurs as a result of the difference in solubility between the small and large crystals [161],

$$\frac{RTY}{M} \ln \frac{S_1}{S_2} = \frac{2\sigma}{\rho}\left(\frac{1}{r_1} - \frac{1}{r_2}\right) \tag{119}$$

where S_1 and S_2 are the solubilities of crystals of radii r_1 and r_2, respectively, σ is the specific surface energy, ρ the density, M the molecular weight of the solute molecules, R the gas constant, and T the absolute temperature. Since r_1 is smaller than r_2, then S_1 is larger than S_2.

Another mechanism for crystal growth is related to polymorphic changes in solutions, and again the driving force is the difference in solubility between the two polymorphs. In other words, the less soluble form grows at the expense of the more soluble phase. This is sometimes also accompanied by changes in the crystal habit. Different faces of the crystal may have different surface energies and deposition may preferentially take place on one of the crystal faces modifying its shape. Other important factors are the presence of crystal dislocations, kinks, surface impurities, and so forth. Most of these effects have been discussed in detail in monographs on crystal growth [162–164].

The growth of crystals in suspension concentrates may create undesirable changes. As a result of the drastic change in particle size distribution, the settling of the particles may be accelerated leading to caking and cementing together of some particles in the sediment. Moreover, increase in particle size may lead to a reduction in biological efficiency. Thus, prevention of crystal growth or at least reducing it to an acceptable level is essential in most suspension concentrates. Surfactants affect crystal growth in a number of ways. The surfactant may affect the rate of dissolution by affecting the rate of transport away from the boundary layer at the crystal solution interface. On the other hand, if the surfactant forms micelles that can solubilize the solute, crystal growth may be enhanced as a result of increasing the concentration gradient. Thus by proper choice of dispersing agent one may reduce crystal growth of suspension concentrates. This has been demonstrated by Tadros [165] for terbacil suspensions. When using Pluronic P75 (polyethylene oxide–polypropylene oxide block copolymer) crystal growth was significant. By replacing the Pluronic surfactant with polyvinyl alcohol the rate of crystal growth was greatly reduced and the suspension concentrate was acceptable.

It should be mentioned that many surfactants and polymers may act as crystal growth inhibitors if they adsorb strongly on the crystal faces, thus preventing solute deposition. However, the choice of an inhibitor is still an art and there are not many rules that can be used for selection of crystal growth inhibitors.

C. Stability Against Claying or Caking

Once a dispersion that is stable in the colloid sense has been prepared, the next task is to eliminate claying or caking. This is the consequence of settling of the colloidally stable suspension particles. The repulsive forces necessary to ensure this colloid stability allows the particles to move past each other forming a dense sediment that is very difficult to redisperse. Such sediments are dilatant (shear thickening, see section on rheology) and hence the SC becomes unusable. Before describing the methods used to control settling and prevention of formation of dilatant clays, an account is given on the settling of suspensions and the effect of increasing the volume fraction of the suspension on the settling rate.

1. Settling of Suspensions

The sedimentation velocity v_o of a very dilute suspension of rigid noninteracting particles with radius a can be determined by equating the gravitational force with the opposing hydrodynamic force as given by Stokes's law, i.e.,

$$\frac{4}{3}\pi a^3 (\rho - \rho_o) g = 6\pi\eta_o a v_o \quad (120)$$

with

$$v_o = \frac{2a^2(\rho - \rho_o)g}{9\eta_o} \quad (121)$$

where ρ is the density of the particles, ρ_o that of the medium, η_o is the viscosity of the medium, and g is the acceleration due to gravity. Equation (121) predicts a sedimentation rate for particles with radius 1 μm in a medium with a density difference of 0.2 g cm^{-3} and a viscosity of 1 mPa (i.e., water at 20°C) of 4.4×10^{-7} m s^{-1}. Such particles will sediment to the bottom of a 0.1-m container in about 60 h. For 10-μm particles, the sedimentation velocity is 4.4×10^{-5} m s^{-1} and such particles will sediment to the bottom of a 0.1-m container in about 40 min.

The above treatment using Stokes's law applies to very dilute suspensions (volume fraction $\varphi < 0.01$). For more concentrated suspensions, the particles no longer sediment independent of each other and one has to take into account both the hydrodynamic interaction between the particles (which applies to moderately concentrated suspensions) and other higher order interactions at relatively high volume fractions. A theoretical relationship between the sedimentation velocity v of nonflocculated suspensions and particle volume fraction has been derived by Maude and Whitmore [166] and by Batchelor [167]. Such theories apply to relatively low volume fractions (< 0.1) and they show that the sedimentation velocity v at a volume fraction φ is related to that at infinite dilution v_o (the Stokes velocity) by an equation of the form:

$$v = v_o (1 - k\varphi) \tag{122}$$

where k is a constant in the region of 5–6. Batchelor [167] derived a rigorous theory for sedimentation in a relatively dilute dispersion of spheres. He considered the reduction in the Stokes velocity with increase in particle number concentration to arise from hydrodynamic interactions. The value of k in Eq. (122) was calculated and found to be 6.55. This theory applies up to a volume fraction of 0.1. At higher volume fractions, the sedimentation velocity becomes a complex function of φ and only empirical equations are available to describe the variation of v with φ. For example, Reed and Anderson [168] developed a virial expansion technique to describe the settling rate of concentrated suspensions. They derived the following expression for the average velocity, v_{av}:

$$v_{av} = v_o \frac{1 - 1.83\varphi}{1 + 4.70\varphi} \tag{123}$$

Good agreement between experimental settling rates and those calculated using Eq. (123) was obtained up to $\varphi = 0.4$.

More recently, Buscall at al. [169] measured the rate of settling of polystyrene latex particles with $a = 1.55$ μm in 10^{-3} mol dm^{-3} up to $\varphi = 0.5$. The results are shown in Figure 6.2. It can be seen that v/v_o decreases exponentially with increase in φ, approaching zero at $\varphi > 0.5$, i.e., in the region of close packing. An empirical equation for the relative settling rate has been derived using the Dougherty–Krieger equation [170] for the relative viscosity, η_r $(= \eta/\eta_o)$,

$$\eta_r = \left(1 - \frac{\varphi}{\varphi_p}\right)^{-[\eta]\,\varphi_p} \tag{124}$$

where $[\eta]$ is the intrinsic viscosity (equal to 2.5 for hard spheres) and φ_p is the maximum packing fraction (which is close to 0.6). Assuming that

$$\frac{v}{v_o} = \alpha \frac{\eta_o}{\eta}$$

it is easy to derive the following empirical relationship for the relative sedimentation velocity, v_r $(= v/v_o)$:

$$\begin{aligned} v_r &= \left(1 - \frac{\varphi}{\varphi_p}\right)^{\alpha [\eta] \varphi_p} \\ &= \left(1 - \frac{\varphi}{\varphi_p}\right)^{k \varphi_p} \end{aligned} \tag{125}$$

By allowing the latex to settle completely and then determining the volume concentration of the packed bed, a value of 0.58 was obtained for φ_p (close to random packing). Using this value and $k = 5.4$, the relative rate of sedimentation was calculated and this gave the full line shown in Figure 6.3, whereas the circles represent the experimental points. Thus, agreement between the calculations using the empirical equation and the experimental results is reasonable.

It seems from the above discussion that there is a correlation between the reduction in sedimentation rate and the increase in relative viscosity of the suspension as the volume fraction of the suspension is increased. This is schematically shown in Figure 6.4 which shows that $v \rightarrow 0$ and $\eta_r \rightarrow \infty$ as $\varphi \rightarrow \varphi_p$. This implies that suspension concentrates with volume fractions approaching the maximum packing do not show any appreciable settling. However, such dense suspensions have extremely high viscosities and are not a practical solution for reduction of settling. In most cases one prepares a suspension concentrate at practical volume fractions (0.2–0.4) and then uses an antisettling agent to reduce settling. As we will discuss in the next section, most of the antisettling agents used in practice are high molecular

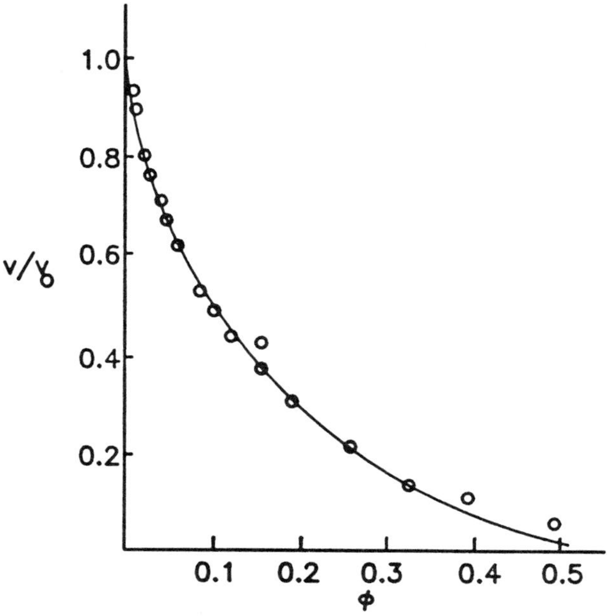

Figure 6.3 Relative sedimentation rate vs. volume fraction for polystyrene latex suspensions.

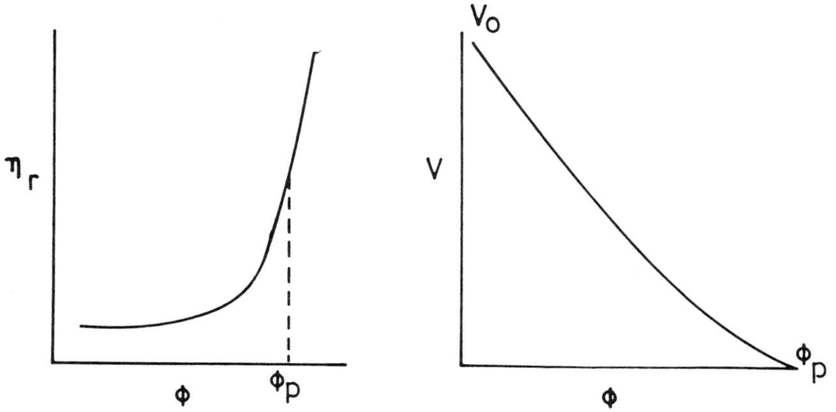

Figure 6.4 Correlation in the reduction in settling rate with the increase in relative viscosity as the volume fraction is increased.

weight polymers. These materials show an increase in the viscosity of the medium with increase in their concentration. However, at a critical polymer concentration (which depends on the nature of the polymer and its molecular weight) they show a very rapid increase in viscosity with further increase in their concentration. This critical concentration (sometimes denoted by C^*) represents the situation where the polymer coils or rods begin to overlap. Under these conditions the solutions become significantly non-Newtonian (viscoelastic; see section on rheology) and produce stresses that are sufficient to overcome the stress exerted by the particles. The settling of suspensions in these non-Newtonian fluids is not simple since one has to consider the non-Newtonian behavior of these polymer solutions. This problem has been addressed by Buscall et al. [169]. In order to adequately describe the settling of particles in non-Newtonian fluids one needs to know how the viscosity of the medium changes with shear rate or shear stress. Most of these viscoelastic fluids show a gradual increase of viscosity with decrease of shear rate or shear stress, but below a critical stress or shear rate they show a Newtonian region with a limiting high viscosity that is denoted as the residual (or zero shear) viscosity. This is illustrated in Figure 6.5 which shows the variation of the viscosity with shear stress for a number of solutions of ethyl hydroxyethylcellulose (EHEC) at various concentrations. It can be seen that the viscosity increases with decrease of stress and the limiting value, i.e., the residual viscosity $\eta(o)$, increases rapidly with increase in polymer concentration. The shear thinning behavior of these polymer solutions is clearly shown, since above a critical stress value the viscosity decreases rapidly with increase in shear stress. The limiting value of the viscosity is reached at low stresses (< 0.2 Pa).

It is now important to calculate the stress exerted by the particles. This stress is equal to $a \Delta\rho g/3$. For polystyrene latex particles with radius 1.55 μm and density 1.05 g cm^{-3}, this stress is equal to 1.6×10^{-4} Pa. Such stress is lower than the critical stress for most EHEC solutions. In this case one would expect a correlation between the settling velocity and the zero shear viscosity. This is illustrated in Figure 6.6 whereby v/a^2 is plotted vs. $\eta(o)$. As is clear, a linear relationship between log (v/a^2) and log $\eta(o)$ is obtained, with a slope

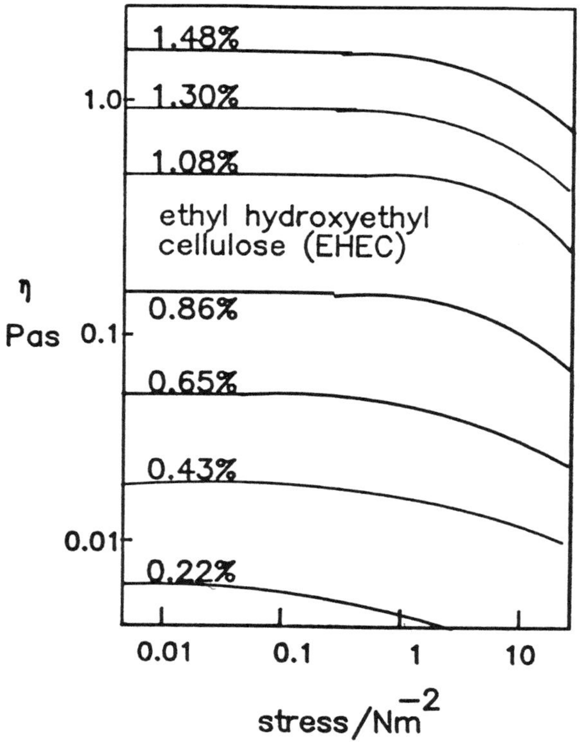

Figure 6.5 Viscosity against shear stress for various concentrations (% w/v) of ethyl hydroxyethylcellulose in water: 0.22; 0.43; 0.65; 0.86; 1.08, 1.30; 1.48.

of −1, over three decades of viscosity. This indicates that the settling rate is proportional to $[\eta(o)]^{-1}$. Thus, the settling rate of isolated spheres in non-Newtonian (pseudoplastic) polymer solutions is determined by the zero shear viscosity in which the particles are suspended. As we will see in the section on rheological measurements, determination of the zero shear viscosity is not straightforward and requires the use of constant stress rheometers.

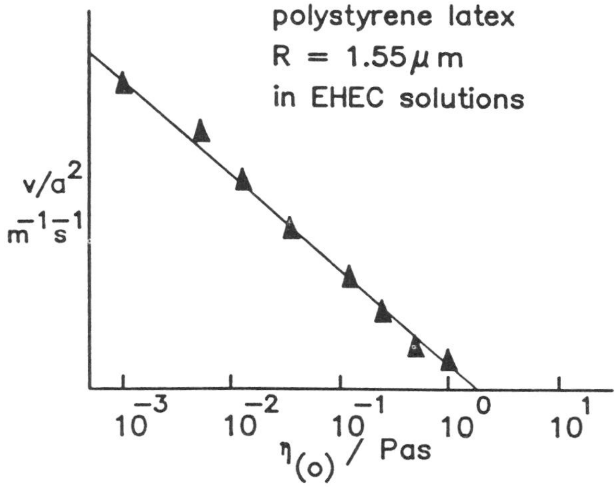

Figure 6.6 v/a^2 against η(o) for a latex with particle radius of 1.55 μm and volume fraction of 0.05 in ethyl hydroxyethylcellulose solutions.

The above correlation applies to the simple case of relatively dilute suspensions. For more concentrated suspensions, other parameters should be taken into consideration, such as the bulk (elastic) modulus. It is also clear that the stress exerted by the particles depends not only on the particle size but on the density difference between the particle and the medium. Many suspension concentrates have particles with radii up to 10 μm and density difference of more than 1 g cm^{-3}. However, the stress exerted by such particles will seldom exceed 10^{-2} Pa and most polymer solutions will reach their limiting viscosity value at higher stresses than this value. Thus, in most cases the correlation between settling velocity and zero shear viscosity is justified, at least for relatively dilute systems. For more concentrated suspensions, an elastic network is produced in the system that encompasses the suspension particles as well as the polymer chains. In this case settling of individual particles may be prevented. However, in this case the elastic network may collapse under its own weight and some liquid is squeezed out from between the particles. This is manifested in a clear liquid layer at the top of the suspension, a phenomenon usually referred to as *syneresis*. If such separation is not significant, it may not cause any problem on application since by shaking the container the whole system redisperses. However, significant separation is not acceptable since it becomes difficult to homogenize the system. In addition such extensive separation is cosmetically unacceptable and the formulation rheology should be controlled to reduce such separation to a minimum.

2. Prevention of Settling and Claying

Several methods are applied in practice to control the settling and prevent the formation of dilatant clays; these methods are summarized below.

(a) Balance of the Density of Disperse Phase and Medium

This is obviously the simplest method for retarding settling since, as clear from Eq. (121), if $\rho = \rho_o$ then $v_o = 0$. However, this method is of limited application and can only be applied to systems where the difference in density between the particle and the medium is not too large. For example, with many organic solids with densities between 1.1 and 1.3 g cm^{-3} suspended in water, some soluble substances such as sugar or electrolytes may be added to the continuous phase to increase the density of the medium to a level equal to that of the particles. However, one should be careful that the added substance does not cause any flocculation for the particles. This is particularly the case when using electrolytes, whereby one should avoid any "salting out" materials, which causes the medium to be a poor solvent for the stabilizing chains. It should also be mentioned that density matching can only be achieved at one temperature. Liquids usually have larger thermal expansion coefficients than solids. If, say, the density is matched at room temperature, settling may occur at higher temperatures. Thus, one has to be careful when applying the density matching method, particularly if the formulation is subjected to large temperature changes.

(b) Use of High Molecular Weight Polymers ("Thickeners")

High molecular weight materials such as natural gums, hydroxyethyl-cellulose or synthetic polymers such as PEO may be used to reduce settling of suspension concentrates. The most commonly used material in agrochemical formulations is xanthan gum (produced by converting waste sugar into a high molecular weight material using a micro-organism and sold under the trade names Kelzan and Rhodopol) which is effective at relatively low concentrations (of the order 0.1–0.2% depending on the formulation). As mentioned above, these high molecular weight materials produce viscoelastic solutions above a critical concentration. This viscoelasticity produces sufficient residual viscosity to stop the settling of individual particles. The solutions also give enough elasticity to overcome separation of the suspension. However, one cannot rule out the interaction of these polymers with the suspension particles which may result in "bridging," and hence the role by which such molecules reduce settling and prevent the formation of clays may be complex. To arrive at the optimum concentration and molecular weight of the polymer necessary for prevention of settling and claying, one should study the rheological characteristics of the formulation as a function of the variables of the system such as its volume fraction, concentration and molecular weight of the polymer, and temperature.

(c) Use of "Inert" Fine Particles

It has long been known that fine inorganic materials such as swellable clays and finely divided oxides (silica or alumina) when added to the dispersion medium of coarser suspensions, can eliminate claying or caking. These fine inorganic materials form a three-dimensional network in the continuous medium which by virtue of its elasticity prevents sedimentation and claying. With swellable clays such as sodium montmorillonite, the gel arises from the interaction of the plate-like particles in the medium. The plate-like particles of sodium montmorillonite consist of an octahedral alumina sheet sandwiched between two tetrahedral silica sheets [171]. In the tetrahedral sheets, tetravalent Si may be replaced by trivalent Al, whereas in the octahedral sheet there may be replacement of trivalent Al with divalent Mg, Fe, Cr, or Zn. This replacement is usually referred to as *isomorphic substitution* [171], i.e., an atom of higher valency is replaced by one of lower valency. This results in a deficit of positive charges or an excess of negative charges. Thus, the faces of the clay platelets become negatively charged and these negative charges are compensated by counterions such as Na^+ or Ca^{2+}. As a result a double layer is produced with a constant charge (that is independent of the pH of the solution). However, at the edges of the platelets some disruption of the bonds occurs resulting in the formation of an oxide-like layer, e.g., -Al-OH, which undergoes dissociation giving a negative (-Al-O^-) or positive (–Al–OH_2^+) depending on the pH of the solution. An isoelectric point may be identified for the edges (usually at pH 7–9). This means that the double layer at the edges is different from that at the faces and the surface charges can be positive or negative depending on the pH of the solution. For that reason, van Olphen [171] suggested an edge-to-face association of clay platelets (which he termed the "house-of-cards" structure) and this was assumed to be the driving force for gelation of swellable clays. However, Norrish [172] suggested that clay gelation is caused simply by the interaction of the expanded double layers. This is particularly the case in dilute electrolyte solutions whereby the double-layer thickness can be several orders of magnitude higher than the particle dimensions.

With oxides, such as finely divided silica, gel formation is caused by formation of chain aggregates, which interact forming a three dimensional network that is elastic in nature. Clearly, the formation of such networks depends on the nature and particle size of the silica particles. For effective gelation, one should choose silicas with very small particles and highly solvated surfaces.

(d) Use of Mixtures of Polymers and Finely Divided Solids

Mixtures of polymers such as hydroxyethyl cellulose or xanthan gum with finely divided solids such as sodium montmorillonite or silica offer one of the most robust antisettling systems. By optimizing the ratio of the polymer to the solid particles, one can arrive at the right viscosity and elasticity to reduce settling and separation. Such systems are more shear thinning than the polymer solutions and hence they are more easily dispersed in water on application. The most likely mechanism by which these mixtures produce a viscoelastic network is probably through bridging or depletion flocculation. The polymer–particulate mixtures also show less temperature dependence of viscosity and elasticity than the polymer solutions and hence they ensure the long-term physical stability at high temperatures.

(e) Controlled Flocculation

For systems where the stabilizing mechanism is electrostatic in nature, e.g., those stabilized by surfactants or polyelectrolytes, the energy–distance curve (Figure 6.1a) shows a secondary minimum at larger particle separations. Such minimum can be quite deep (few tens of kT units), particularly for large (> 1 μm) and asymmetrical particles. The depth of the minimum also depends on electrolyte concentration. Thus, by adding small amounts of electrolyte weak flocculation may be obtained. The weakly flocculated may produce a gel network (self-structured systems) of sufficient elasticity to reduce settling and eliminate claying. This has been demonstrated by Tadros [172] for ethirimol suspensions stabilized with phenol formaldehyde sulfonated condensate (a polyelectrolyte with modest molecular weight). The energy–distance curves for such suspensions at three NaCl concentrations (10^{-3}, 10^{-2}, and 10^{-1} mol dm^{-3}) are shown in Figure 6.7. It can be seen that by increasing NaCl concentration, the depth of the secondary minimum increases, reaching about 50 kT at the highest electrolyte concentration. By using electrolytes of higher valency such as $CaCl_2$ or $AlCl_3$, such deep minima are produced at much lower electrolyte concentrations. Thus, by controlling electrolyte concentration and valency, one can reach a sufficiently deep secondary minimum for producing a gel with enough elasticity to reduce settling and eliminate claying. This is illustrated in Figure 6.8 which shows the variation of sediment height and redispersion as a function of electrolyte concentration for three electrolytes: NaCl, $CaCl_2$, and $AlCl_3$. It is clear that above a critical electrolyte concentration, the sediment height

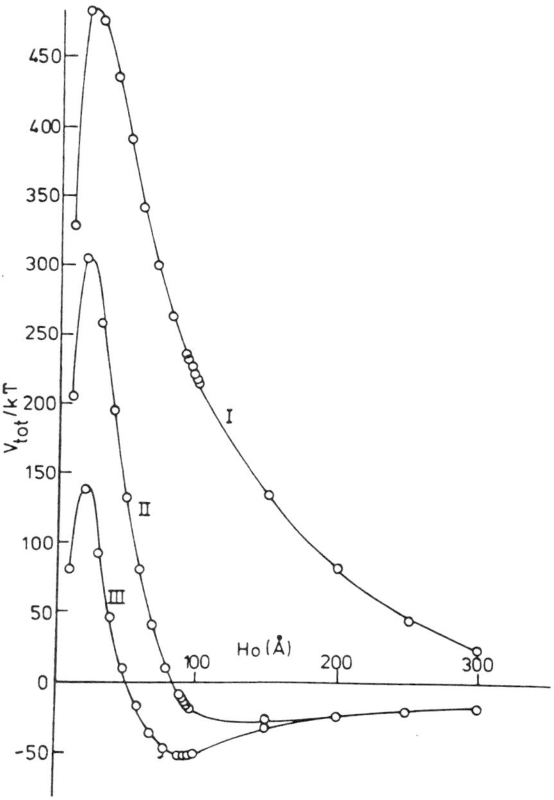

Figure 6.7 Energy–distance curves for ethirimol suspensions at three NaCl concentration: I, 10^{-3}; II, 10^{-2}; III, 10^{-1} mol dm^{-3}.

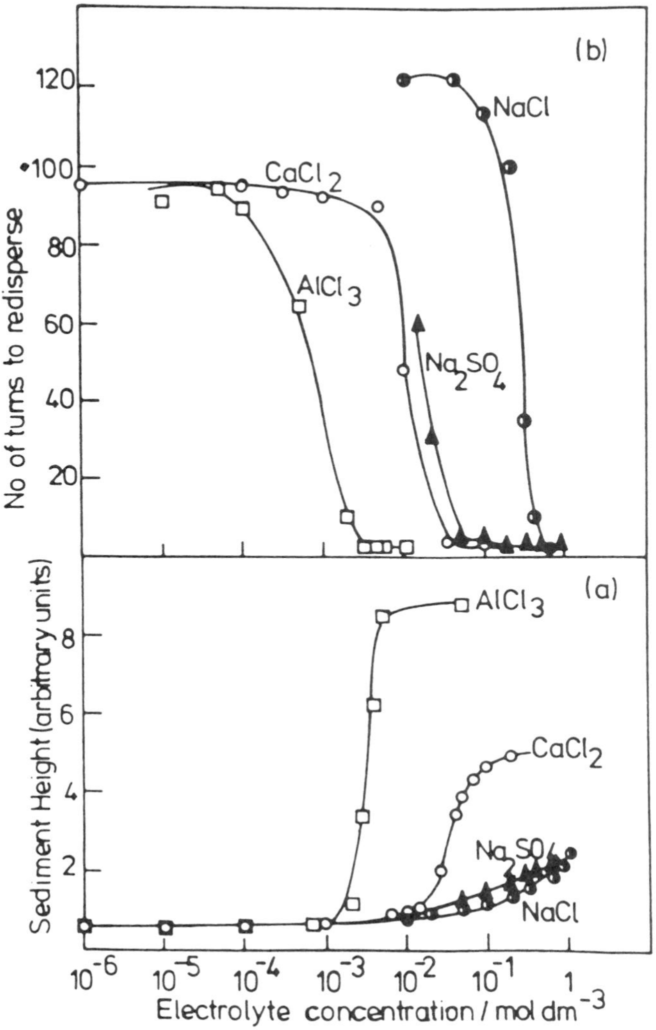

Figure 6.8 Sediment height and redispersion as a function of electrolyte concentration.

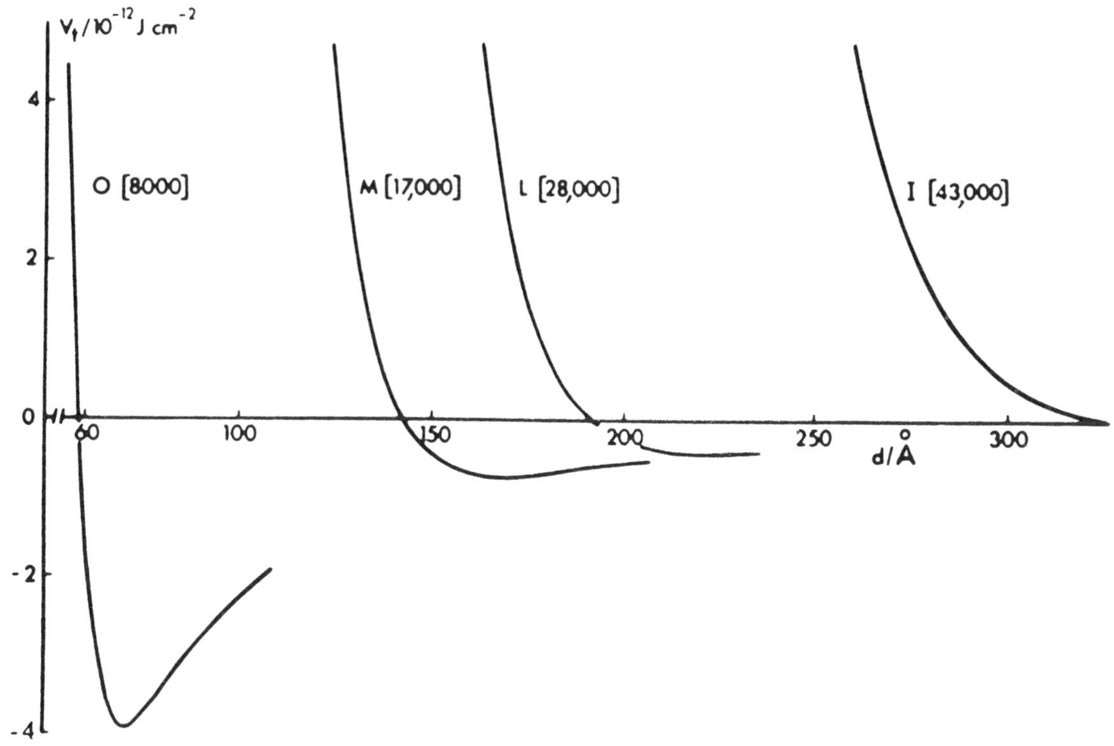

Figure 6.9 Energy–distance curves for polystyrene latex dispersions with adsorbed PVA layers with various molecular weights.

increases and this prevents the formation of clays. Above this critical electrolyte concentration, redispersion of the suspension becomes easier as illustrated in Figure 6.8.

For systems stabilized by nonionic surfactants or macromolecules, the energy–distance curve also shows a minimum (Figure 6.1b) whose depth depends on particle size, Hamaker constant, and thickness of the adsorbed layer [173]. This is illustrated in Figure 6.9 where the energy–distance curves for polystyrene latex particles containing PVA layers of various molecular weights [173]. It is clear from Figure 6.9 that with the high molecular weight polymers ($M > 17{,}000$ with an adsorbed layer thickness $\delta > 9.8$ nm) the energy minimum is too small for flocculation to occur. However, as the molecular weight of the polymer is reduced below a certain value, i.e., as the adsorbed layer become small ($M = 8000$ and $\delta = 3.3$), the energy minimum becomes deep enough for flocculation to occur. This was demonstrated for the latex containing PVA with $M = 8000$, whereby scanning electron micrographs of a freeze-dried sediment showed flocculation and an open structure. In this case claying was prevented.

(f) Depletion Flocculation

The addition of "free" (nonadsorbing) polymer can induce weak flocculation of the suspension, when the concentration or volume fraction of the free polymer (φ_p) exceeds a critical value denoted by φ_p^+. The first quantitative analysis of the phenomenon was reported by Asakura and Oosawa [174]. They showed that when two particles approach to a distance of separation that is smaller than the diameter of the free coil, exclusion of the polymer molecules from the interstices between the particles takes place, leading to the formation of a polymer free zone (depletion zone). This is illustrated in Figure 6.10, which shows the situation below and above φ_p^+. As a result of this process, an attractive force, associated with the lower osmotic pressure in the region between the particles, is produced. This weak flocculation process can be applied to prevent sedimentation and formation of clays. This has been illustrated by Heath et al. [175] using ethirimol suspensions stabilized by a graft copolymer containing PEO side chains (with $M = 750$) to which free PEO with various molecular weights (20,000,

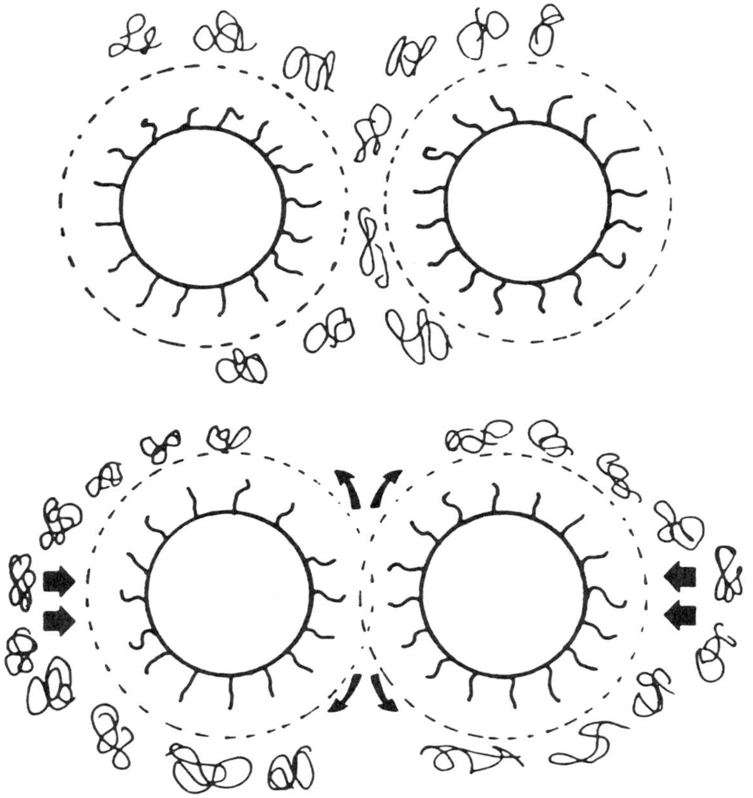

Figure 6.10 Schematic representation of depletion flocculation. (top) below φ_p^+; (bottom) above φ_p^+.

35,000, and 90,000) was added. Above a critical volume fraction of the free polymer (which decreased with increase of molecular weight) weak flocculation occurred and this prevented the formation of dilatant sediments.

IV. ASSESSMENT OF THE LONG-TERM PHYSICAL STABILITY OF SUSPENSION CONCENTRATES

For the full assessment of the properties of suspension concentrates, three main types of measurements are required. First, some information is required on the structure of the solid/solution interface at a molecular level. This requires investigation of the double-layer properties (for systems stabilized by ionic surfactants and polyelectrolytes), adsorption of the surfactant or polymer, as well as the extension of the layer from the interface (adsorbed layer thickness). Second, one needs to obtain information on the state of dispersion on standing, such as its flocculation and crystal growth. This requires measurement of the particle size distribution as a function of time and microscopic investigation of flocculation. The spontaneity of dispersion on dilution, i.e., reversibility of flocculation also needs to be assessed. Finally, information on the bulk properties of the suspension on standing is required and this could be obtained using rheological measurements. Below is given a brief description of the methods that may be applied for suspension concentrates.

A. Double-Layer Investigation

The most practical method for investigating the double layer at the solid/solution interface is through electrokinetic measurements [176]. The most common electrokinetic measurement is that of microelectrophoresis, which allows one to obtain the particle mobility as a function of the parameters of the system such as surfactant and electrolyte concentration. The dilute dispersion is investigated microscopically in a capillary tube connected to two containers fitted with electrodes. By applying an electric field with a strength E/l, where E is the applied voltage and l the distance between the electrodes, one can measure the average velocity v of the particles and hence its mobility u ($u = \mathrm{v}/(E/l)$). From the mobility u, the ζ potential can be calculated using the Smoluchowski equation, which is valid for most coarse suspensions [176]:

$$u = \frac{\varepsilon \varepsilon_o \zeta}{\eta} \tag{126}$$

where ε is the relative permittivity, ε_o the permittivity of free space, and η the viscosity of the medium. For aqueous dispersions at 25°C,

$$\zeta = 1.282 \times 10^6 u \tag{127}$$

ζ is calculated in volts when u is expressed in $m^2\ V^{-1}\ s^{-1}$.

By measuring the ζ as a function of concentration of added ionic surfactant, one obtains the information on limiting ζ potential that can be reached. This value usually coincides with the saturation adsorption. Qualitatively, the higher the ζ potential, the stronger the repulsion between the particles and the higher its colloid stability.

B. Surfactant and Polymer Adsorption

The adsorption of ionic, nonionic, and polymeric surfactant on the agrochemical solid gives valuable information on the magnitude and strength of the interaction between the molecules and the substrate as well as the orientation of the molecules. The latter is important in determining the colloid stability. Adsorption isotherm determination is fairly simple but it requires careful experimental techniques. A representative sample of the solid with known surface area A per unit mass must be available. The surface area is usually determined using gas adsorption. N_2 is usually used as the adsorbate, but for materials with relatively low surface area, such as those encountered with most agrochemical solids, it is preferable to use Kr as the adsorbate. The surface area is obtained from the amount of gas adsorbed at various relative pressures by application of the BET equation [177]. However, the surface area determined by gas adsorption may not represent the true surface area of the solid in suspension (the so called wet surface). In this case it is preferable to use dye adsorption for measurement of the surface area [178].

An analytical method that is sensitive enough for determination of surfactant or polymer adsorption needs to be established. Several spectroscopic and colorimetric methods may be applied, which in some cases must be developed for a particular surfactant or polymer. A reproducible method for dispersing the solid in the surfactant or polymer solution needs also to be established. In this case high speed stirrers or ultrasonic radiation may be applied provided these do not cause any comminution of the particles during dispersion. The time required for adsorption must be established by carrying out experiments as a function of time. Finally, a procedure for separation of the solid from the solid must be established. This may be carried out using centrifugation.

Once the above procedure is established, the adsorption isotherm can be determined, whereby the amount of adsorption in moles per unit area, Γ, is plotted as function of equilibrium concentration, C_2. With many surfactants, where the adsorption is reversible, the results can usually be fitted using a Langmuir-type equation, i.e., Γ increases gradually with increase of C_2 and eventually reaches a plateau value, Γ_∞, which corresponds to saturation adsorption:

$$\Gamma = \frac{\Gamma_\infty b C_2}{1 + bC_2} \tag{128}$$

where b is a constant that is related to the adsorption free energy

$$b \propto \exp - \frac{\Delta G_{ads}}{RT}$$

A linearlized form of the Langmuir equation may be used to obtain Γ_∞ and b, i.e.,

$$\frac{1}{\Gamma} = \frac{1}{\Gamma_\infty} + \frac{1}{\Gamma_\infty b C_2} \tag{129}$$

A plot of $1/\Gamma$ vs. $1/C_2$ gives a straight line with intercept $1/\Gamma_\infty$ and slope $1/\Gamma_\infty bC_2$ from which both Γ_∞ and b can be calculated.

As discussed in Chapter 3, from Γ_∞, the area per surfactant molecule can be calculated (area per molecule = $1/\Gamma_\infty N_{av}$). The area per surfactant molecule gives information on the orientation at the solid/solution interface. For example, for vertical orientation of ionic surfactant molecules such as sodium dodecyl sulfate on a hydrophobic surface, an area per surfactant ion in the region of 0.3–0.4 nm^2 is to be expected. The area per surfactant ion is determined by the cross sectional area of the ionic head group. However, many surfactant ions may undergo association on the solid surface and this is usually accompanied by steps in the adsorption isotherm [179]. In the region of surfactant association, the amount of adsorption increases rapidly with increase in surfactant concentration and finally another plateau is reached when the aggregate units (sometimes referred to as hemimicelles) become close packed on the surface.

With nonionic surfactants, the adsorption isotherm may also show steps that are characteristic of various orientation and association of the molecules on the surface [180]. Nonionic surfactants of the ethoxylate type such as $R-(CH_2-CH_2-O)_n-OH$ show complex adsorption isotherms that are very sensitive to small changes in concentration, temperature, or molecular structure. The main interaction with a hydrophobic surface is usually through hydrophobic bonding (see Chapter 3) with the alkyl group, leaving the PEO chain dangling in solution. As a result, the area per molecule is usually large since it is determined by the area occupied by the PEO chain that occupies a large area, particularly when the chain contains several EO units. As with the case of ionic surfactants, the molecules may aggregate on the surface forming hemimicelles or even micelles. A schematic representation of nonionic surfactant molecules adsorbed on a solid surface was given in Chapter 3 which shows the possible association structures. Moreover, the adsorption increases rapidly with increase of temperature near the phase separation point.

Polymer adsorption is more complex than surfactant adsorption, since one must consider the various interactions (chain-surface, chain/solvent and surface/solvent) as well as the conformation of the polymer on the surface [181]. The various conformation of different macromolecular surfactants was schematically shown in Chapter 3.

Complete information on polymer adsorption may be obtained if one is able to determine the segment density distribution, i.e., the segment concentration in all layers parallel to the surface where such segments are accommodated [182]. In practice, however, such information is unavailable and therefore one determines three main parameters: the amount of adsorption per unit area, Γ, the fraction of segments in trains p, and the adsorbed layer thickness δ. Γ can be determined in the same way as for surfactants, although in this case some complications may arise. First, adsorption may be very slow requiring long equilibration times (sometimes days). Second, most commercially available polymers have a wide distribution of molecular weights. Instead of the theoretically predicted high-affinity isotherms, in which the plateau value starts at near-zero polymer concentration, the experimental isotherm is rounded in nature. The rounded isotherms are due to the heterodispersity of the samples [183]. In this case, the amount adsorbed Γ depends on the area to volume ratio A/V, with Γ decreasing as A/V increases (i.e., the amount of adsorption decreases with concentration of the suspension).

The second parameter that needs to be established in polymer adsorption is the fraction p of segments in direct contact with the surface. As mentioned in Chapter 3, direct and indirect methods may be applied. The direct methods are based on spectroscopic techniques [181] such as infrared (IR), electron spin resonance (ESR), electron paramagnetic resonance (EPR), and nuclear magnetic resonance (NMR) [184, 185]. Of the indirect methods, microcalorimetry is perhaps the most convenient to apply. By measuring the heat of adsorption of the chain, one may obtain p by referring to the heat of adsorption of a segment.

The most practical methods for measuring the adsorbed layer thickness are based on hydrodynamic techniques (see Chapter 3). Unfortunately, these methods can only be applied to spherical particles with small radii such that the ratio of the adsorbed layer thickness to particle radius δ/a is significant (of the order of 10%). The adsorbed layer thickness is determined from a comparison of the hydrodynamic radius of the particle with the adsorbed layer a_δ with that of the bare particle

a. Thus, measurement of the adsorbed layer thickness on agrochemical suspension particles is not possible. One has to use model particles such as polystyrene latex particles to obtain such information and the assumption is made that the adsorbed layer thickness obtained on such model particles is comparable to that on the practical agrochemical suspension particles. This assumption is not serious, since in most cases one only needs a comparison between various polymer samples. Four different hydrodynamic methods may be applied to obtain δ: measurement of the sedimentation coefficient (using an ultracentrifuge), diffusion coefficient (using dynamic light scattering or photon correlation spectroscopy), viscosity, and slow-speed centrifugation. These methods were described in some detail in Chapter 3.

C. Assessment of the State of the Dispersion

Two general techniques are widely used for monitoring the flocculation rate of suspensions, both of which can only be applied to dilute systems. The first method is based on measurement of the turbidity τ (at a given wavelength of light λ) as a function of time during the early stages of flocculation. This method can be only applied if the particle size of the particles is smaller than $\lambda/20$ and hence it cannot be used for coarse suspensions. In the latter case, direct particle counting as a function of time is the most suitable procedure. This can be carried out manually using a light microscope or automatically using a Coulter counter or an ultramicroscope. Recently, optical microscopy was combined with image analysis techniques for counting the particles. The rate constant for flocculation (assumed to follow a bimolecular process) is determined by plotting $1/n$ vs. t, where n is the particle number at time t, i.e.,

$$\frac{1}{n} = \frac{1}{n_o} + kt \qquad (130)$$

where n_o is the number of particles at $t = 0$. The rate constant k can be related to the rapid flocculation rate k_o given by Smoluchowski [186], i.e.,

$$k_o = \frac{8kT}{6\eta} \tag{131}$$

For particles dispersed in an aqueous phase at 25°C, $k_o = 5.5 \times 10^{-18}$ $m^3\ s^{-1}$. k is usually related to k_o by the stability ratio W, i.e.,

$$W = \frac{k_o}{k}$$

The higher the value of W, the more stable the dispersion. Thus, by plotting W vs. the parameters of the system such as surfactant and/or electrolyte concentration, one can obtain a quantitative assessment of the stability of the suspension under various conditions. It should be mentioned that the stability ratio W is related to the energy maximum in the energy–distance curve for electrostatically stabilized suspensions. The higher this energy maximum, the higher the value of W.

Incipient flocculation of sterically stabilized suspensions, i.e., the condition when the chains are in poor solvent condition, can be investigated using turbidity measurement. The suspension is placed in a spectrophotometer cell placed in a block that can be heated at a controlled rate. From a plot of turbidity vs. temperature one can obtain the critical flocculation temperature, which is the point at which there is a rapid increase in turbidity.

Crystal growth (Ostwald ripening) can be investigated by following the particle size distribution as a function of time using a Coulter counter or an optical disk centrifuge. The percentage number cumulative frequency over size is plotted vs. time for various particle radii. Curves are produced at various intervals of time. When crystal growth occurs, the cumulative counts are shifted toward coarser particle sizes. Horizontal lines corresponding to various percentage cumulative counts are then made to cut the curves in their steep portions. From the intersections, a plot of equivalent diameter against time is drawn, allowing one to obtain the rate of crystal growth.

A qualitative method for assessment of the state of dispersion is that of sediment height (volume) measurement. The suspension is placed in graduated stoppered cylinders and the sediment height (volume) is followed as a function of time until equilibrium is reached. Normally in sediment height (volume) measurements one compares the initial height H_o or volume V_o with that reached at equilibrium H and V, respectively. A clayed suspension gives low values for the relative height (H/H_o) or volume (V/V_o), whereas a "structured" suspension containing an antisettling agent or weakly flocculated gives high values. Clearly the higher the value of H/H_o or V/V_o the better the suspension. One aims at a relative value of unity which implies no separation on standing. However, one must be careful since strong flocculation must be avoided. A strongly flocculated system may give a high relative sediment height (volume) but in this case the suspension cannot be redispersed or adequately diluted. Thus, at the end of the sedimentation experiment, one should redisperse the system by rotating the cylinders end-over-end and carry out a dispersion test. The suspension is poured into a beaker containing water and the dispersion observed visually. In most application methods one requires a spontaneous dispersion on dilution with minimum agitation. For a more quantitative assessment of the state of flocculation and dispersion of the suspension concentrate, one should apply rheological methods as discussed below.

D. Rheological Measurements

Rheology deals with the deformation of matter including flow [187]. Basically, one needs to establish three parameters: shear stress τ (force per unit area, N m^{-2} or Pa), strain γ (the relative deformation which is dimensionless), and shear rate $\dot{\gamma}$ (= $d\gamma/dt$, i.e., s^{-1}). The ratio of the stress and strain τ/γ gives the shear modulus G (Pa), whereas the ratio of the stress and shear rate $\tau/\dot{\gamma}$ gives the viscosity η (Pa).

Typical plots of τ vs. $\dot{\gamma}$ are shown in Figure 6.11. The flow curves can be described by equations that describe the variation of stress with shear rate. For Newtonian systems (curve a), such as those encountered with very dilute stable suspensions, the stress is related to the shear rate by the following simple equation:

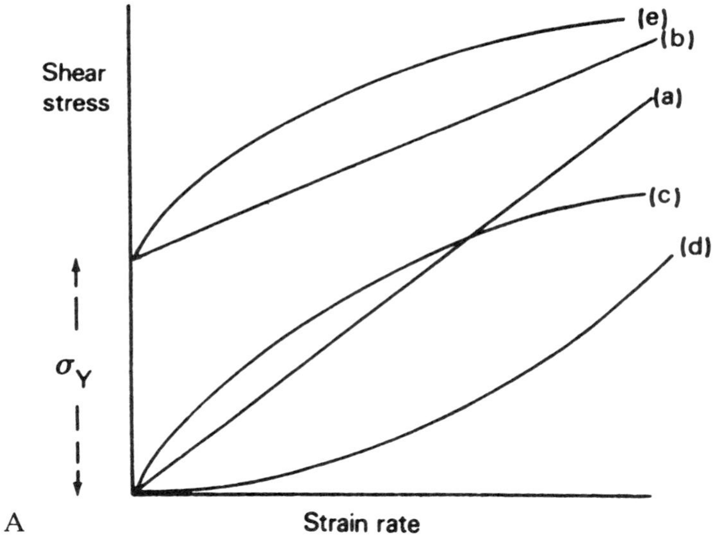

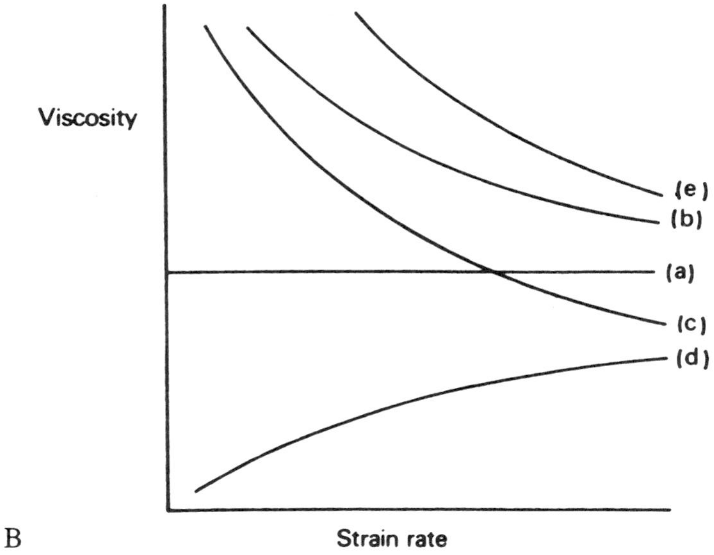

Figure 6.11 (A) Flow curves for five systems: (a) Newtonian; (b) Bingham plastic; (c) pseudoplastic (shear thinning); (d) Dilatant (shear thickening); (e) Pseudoplastic with a yield stress. (B) Corresponding curves for the variation of η with $\dot{\gamma}$.

$$\tau = \eta\dot{\gamma} \tag{132}$$

Equation (132) shows that η is independent of the applied shear rate as indicated in Figure 6.11B. However, with most suspension concentrates, particularly those that are "structured" or weakly flocculated, the stress-shear rate curves are shown by curves (b) or (c). Curve (b) is a typical Bingham plastic system with a well-defined yield stress (the intercept on the y axis) and the rheological equation is

$$\tau = \tau_{\beta} + \eta_{pl}\,\dot{\gamma} \tag{133}$$

η_{pl} is the slope of the linear curve and is termed the *plastic viscosity.* The most common flow curve for "structured" or weakly flocculated suspension concentrate is that represented by (c),which is termed *shear thinning* or *pseudoplastic.* In this case the viscosity decreases with increase of shear rate as shown in Figure 6.11B. The flow curve can be represented by a power law of the form:

$$\tau = \eta\,\dot{\gamma}^{n} \tag{134}$$

with $n < l$. As we will see later, many psuedoplastic systems show a limiting high Newtonian viscosity below a critical shear rate (or shear stress), termed the *residual* (*zero shear*) *viscosity.*

Many stable concentrated suspensions (which are not structured or weakly flocculated) show a stress–shear rate relationship represented by curve (d), which implies that the viscosity increases with increase of applied shear rate, i.e., shear thickening or dilatant systems. As discussed before, the sediments in such stable suspensions show this shear thickening behavior (clays or cakes). Such flow behavior has to be avoided in practice since the clayed suspension cannot be redispersed by shaking. The flow curve can also be represented by a power law as given in Eq. (134) but with $n > 1$.

Another possible flow curve is that represented by curve (e) which is a shear thinning system but with a yield stress. The flow curve can be represented by the Herschel–Buckley equation [187], i.e.,

$$\tau = \tau_\beta + \eta\dot{\gamma}^n \tag{135}$$

with $n < 1$.

A semiempirical linear relationship between stress and shear rate was derived by Casson [188], i.e.,

$$\tau^{1/2} = \tau_c^{1/2} + \eta_c^{1/2}\dot{\gamma}^{1/2} \tag{136}$$

The above equation shows that a plot of $\tau^{1/2}$ vs. $\dot{\gamma}^{1/2}$ gives a straight line with a slope equal to $\eta_c^{1/2}$ and an intercept equal to $\tau_c^{1/2}$ (the Casson yield stress). It should be mentioned that the Casson yield stress is not equal to the extrapolated Bingham yield stress that is usually obtained by extrapolation of the linear portion of the pseudoplastic curve to $\dot{\gamma} = 0$. In most cases $\tau_c < \tau_\beta$.

In the above analysis, the assumption is made that a steady state is reached and the system does not show any time effects. However, in many cases the system may not reach a steady state during the rheological measurement as a result of the change of the structure of the suspension during shearing. For example, some aggregates may be broken under shear and/or some aligning of irregularly shaped particles may take place. This results in a reversible time-dependent decrease of viscosity, usually referred to as thixotropy. This may be illustrated if a sequence of shear stress–shear rate measurements are carried out, whereby the shear stress is gradually increased to a maximum defined shear rate and then decreased from the highest to the lowest value in a reverse sequence. If the system shows time dependence, the ascending and descending curves do not coincide and shows instead a hysteresis loop. This effect results from the fact that any structure breakdown and/or alignment of the particles does not reverse during the period within which the rheological experiment is carried out. Such systems show a continuous reduction in the viscosity at a given shear rate with the time of shear. When the shear is removed the system recovers its initial viscosity within a set period of time that depends on the system. Clearly, thixotropy is undesirable in agrochemical suspension concentrates. During transportation, the suspension is subjected to shearing forces, which may result in breakdown of the weak structure that is built to overcome settling and claying. If such a structure does not build up within a short time, the suspension may undergo settling and claying during transportation. Thus, in practice a shear thinning system is desirable with as little time dependence (thixotropy) as possible. In addition, the limiting high shear viscosity reached should not be too high (usually less than 50–100 mPa), otherwise redispersion and spontaneity of dispersion on dilution becomes difficult. However, the limiting low shear viscosity should be made as high as possible to overcome settling and claying.

For the practical measurement of the steady state shear stress–shear rate curves one uses rotational viscometers fitted with concentric cylinder or cone and plate geometries. The suspension is placed between the gap of the concentric cylinders or the cone and plate. The gap width should be at least 10 times larger than the maximum particle size in the suspension. The inner or outer cylinder of the concentric cylinder viscometer is rotated with an angular velocity Ω and the torque T on the other cylinder is measured. The shear rate is calculated from the angular velocity and the stress from the torque.

The above measurements are usually referred to as high-deformation measurements since the experiments are carried out under conditions where the structure of the system is broken down during the flow. Although these measurements give valuable information on the behavior of the system under application, such as redispersion, they do not give enough information on the system on standing (storage). For the latter purpose, low-deformation (i.e., under conditions whereby the structure of the system is maintained) measurements are required. Two main types of such measurements are available, namely, constant stress (creep) and dynamic (oscillatory) measurements. In creep measurements, a constant stress τ is applied to the system and the strain or deformation is measured as a function of time. The results are usually expressed as creep compliance J (strain per unit applied stress, i.e., $m^2\ N^{-1}$) vs. t. As an illustration Figure 6.12 shows typical creep curves for a pesticidal suspension concentrate (containing 250 g dm^{-3}) structured with bentonite clay [189] at 30 or 45 g cm^{-3}. The creep curves are characterized by three main regions. Directly after the

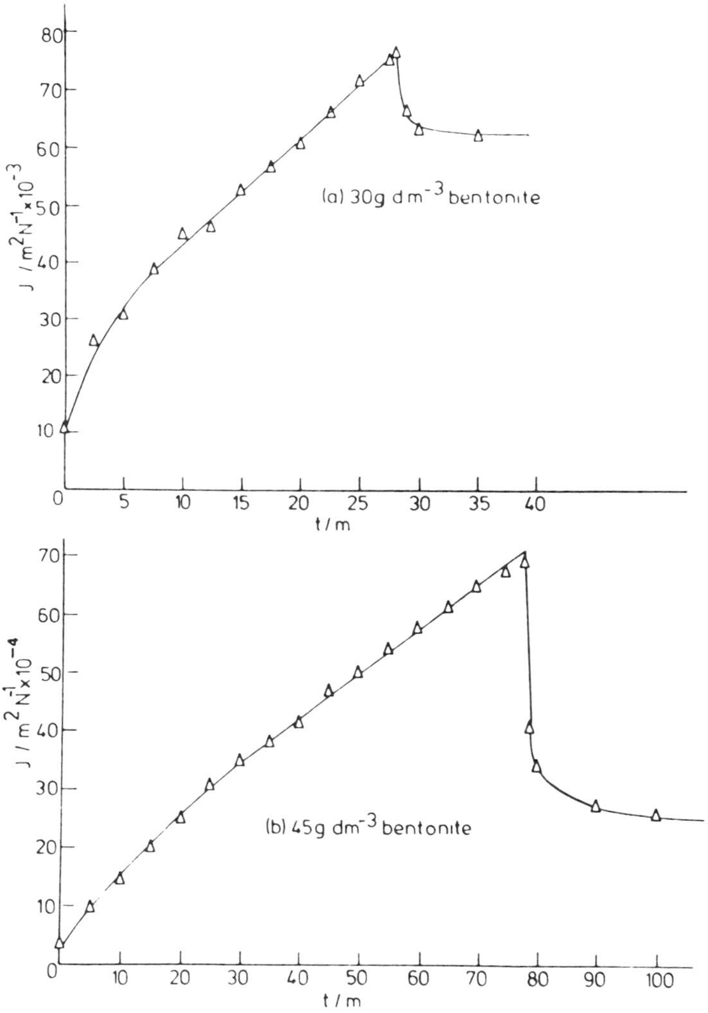

Figure 6.12 Creep curves for a pesticidal suspension concentrate (250 g dm^{-3}) at two bentonite concentrations.

application of the stress, one observes a rapid elastic deformation resulting in an instantaneous compliance J_o, i.e., an instantaneous shear modulus $G_o = 1/J_o$. After this instantaneous response, the flow curve shows a slow elastic (retarded) deformation characterized by a slower increase in compliance with increase of time. In this region, bonds are broken and reformed, but not at the same rate and a spectrum of retarded elastic compliances is obtained. Finally, after some time that depends on the system, a steady state is reached whereby the compliance shows linear increase with time. In other words, a purely viscous response is produced. The slope of this linear region gives the viscosity at the applied stress, η_τ. If the stress is removed at the time when such steady state is reached, the creep curve reverses sign but only the elastic component of the system may be recovered.

The above creep curves enable one to obtain two main parameters, namely, G_o and η_τ. In the above example of suspension concentrate stabilized by bentonite, a value of G_o of 93 Pa is obtained for the system containing 30 g dm^{-3}. With the suspension containing 45 g dm^{-3} bentonite, G_o is much higher. This is consistent with the lack of separation of the system containing the higher bentonite concentration. If creep curves are carried out at various stresses, a plot of η_τ vs. τ allows one to obtain the residual (zero shear) viscosity. This is illustrated in Figure 6.13. As discussed above, the zero shear viscosity is the parameter that determines the settling of individual particles. Thus, creep measurement can be applied in practice for the assessment and predication of the long term physical stability of suspension concentrates. Unfortunately, such measurements are time consuming and require specially designed constant stress (CS) rheometers. In recent years, many of such CS rheometers became commercially available and they offer valuable tools to the agrochemical formulation chemist. Such instruments are computer controlled allowing one to obtain creep curves as a function of applied stress and the viscosity is automatically computed and plotted vs. the applied stress to obtain the residual viscosity. In addition the results allow one to obtain the critical stress at which flow begins to occur. This critical stress may also be related to the settling and separation of the suspension.

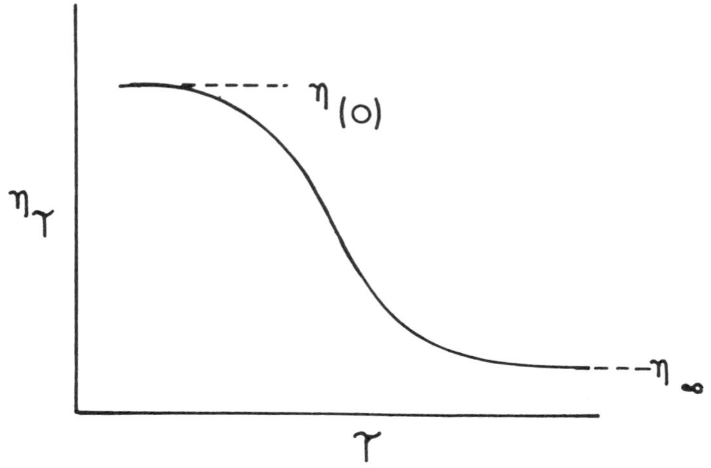

Figure 6.13 Viscosity as a function of applied shear stress.

An alternative low deformation experiment that also gives valuable information on the suspension concentrate on standing is the dynamic or oscillatory technique [190]. In these measurements, one applies a sinusoidal strain (or stress) with a frequency ν in Hz or ω in rad s^{-1} ($\omega = 2\pi\nu$) on the system (which may be placed in the gap between concentric cylinders or a cone and plate geometry). This is illustrated in Figure 6.14. The strain and stress are simultaneously measured to obtain their amplitudes and the time shift, Δt, between the sine waves of strain and stress. The latter shift occurs for a viscoelastic material as is the case with suspension concentrates. The product of Δt and ω gives the phase angle shift δ in radians. Note that for a purely elastic system $\delta = 0$, for a purely viscous system $\delta = 90°$, and for a viscoelastic system $0 < \delta < 90°$.

The ratio of the stress amplitude, τ_o, to the strain amplitude, γ_o, gives the complex modulus, G^*, i.e.,

$$|G^*| = \frac{\tau_o}{\gamma_o} \tag{137}$$

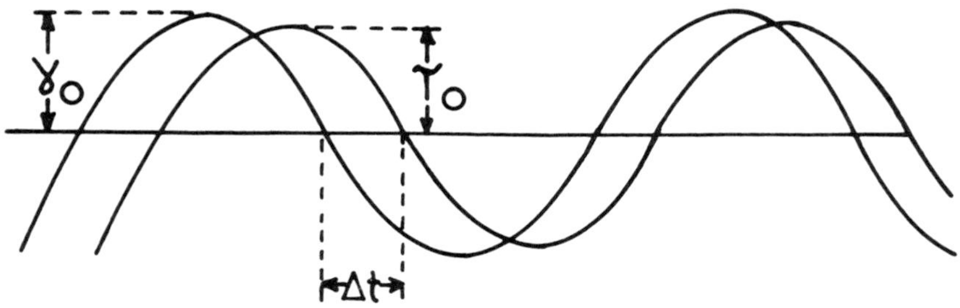

Figure 6.14 Corresponding stress and strain waveforms for a viscoelastic system.

The complex modulus is formed from a storage component G' (which is a measure of the elastic energy stored in a cycle) and a loss component G'' (which is a measure of the energy dissipated as viscous flow in a cycle). G' and G'' are related to G^* by the following equations:

$$G' = |G^*| \cos \delta \tag{138}$$

$$G'' = |G^*| \sin \delta \tag{139}$$

and,

$$|G^*| = G' + iG'' \tag{140}$$

$$\frac{G''}{G'} = \tan \delta \tag{141}$$

The dynamic viscosity η' is also obtained from G'' and ω:

$$\eta' = \frac{G''}{\omega} \tag{142}$$

It should be mentioned that η' approaches the zero shear viscosity $\eta(o)$ in the limit $\omega \rightarrow 0$.

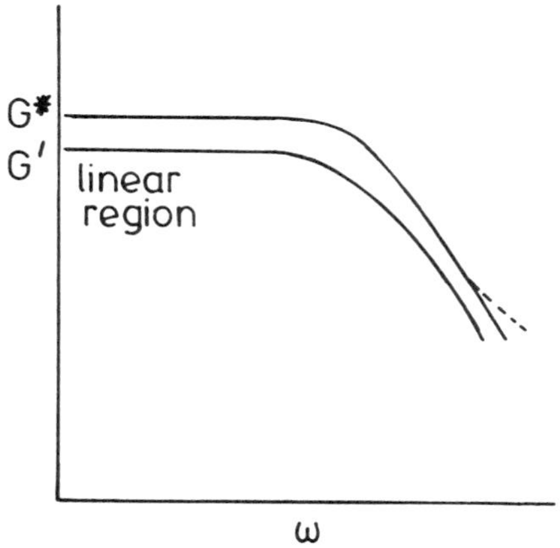

Figure 6.15 Variation of G^* and G' with strain amplitude.

To carry out dynamic measurements (e.g., using a Bohlin VOR, Bohlin Reologie, Lund, Sweden), the suspension is placed in the gap between the concentric cylinders. A sinusoidal strain is applied to the cup, while the stress on the bob is monitored using a torque element attached to this bob. The relative displacement is measured using circular transducers. Measurements are usually carried out at constant frequency by gradually increasing the strain amplitude from the lowest possible value at which a measurement can be made. The resulting moduli values and the phase angle shift are plotted vs. the relative strain to identify the linear viscoelastic region, where the moduli values are independent of the magnitude applied strain amplitude. This is illustrated in Figure 6.15 which shows the region whereby both G^* and G' are constant (the linear viscoelastic region). Above a critical strain amplitude, γ_{cr}, both G^* and G' begin to decrease (whereas G'' begins to increase). Once the viscoelastic region is established, measurements are then made at constant strain (within this region) while varying the frequency ν. The latter is usually varied by three to four decades depending on the instrument range. The rheological parameters, G^*, G', G'', and η', are plotted as a function of frequency. This is

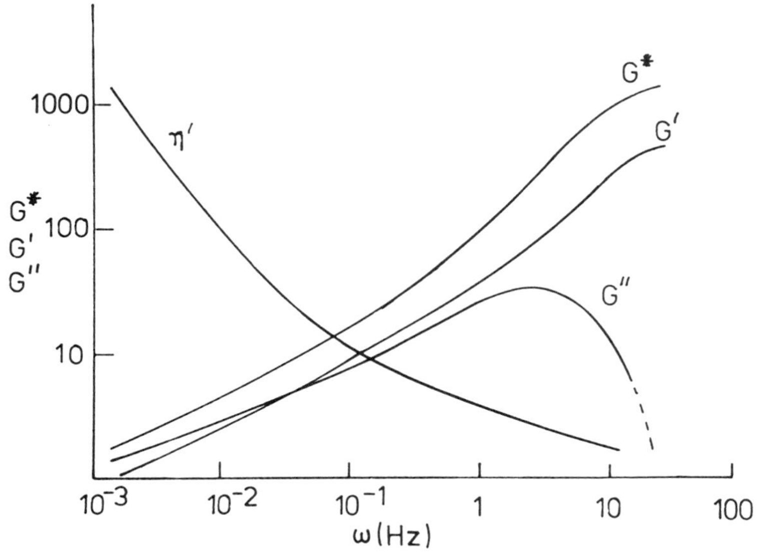

Figure 6.16 Schematic representation of the variation of G^*, G', G'', and η' with frequency.

illustrated in Figure 6.16. The trend obtained, which depends on the system, can be used to obtain information on the "structure" of the suspension. Many structured and weakly flocculated suspensions show the behavior illustrated in Figure 6.16. In the low frequency range ($< 10^{-2}$ Hz), the system is usually more viscous than elastic. In this low frequency regime (which corresponds to long time), the system can dissipate energy as viscous flow. In contrast, in the high frequency regime ($>$ 1 Hz), $G' > G''$, since in this region most of the energy is stored elastically. Indeed at sufficiently high frequency (which depends on the system) G' approaches G^* and G'' tends toward zero.

The above measurements can be used for the control and assessment of the long term physical stability of suspension concentrates. One usually aims at a system that shows a long linear viscoelastic region. This implies that the "gel" structure of the system is sufficiently coherent to prevent extensive separation. Separation may be prevented altogether when the storage modulus (measured at a frequency > 1 Hz) reaches an optimum value. However, one must be careful in the choice of this optimum modulus since high values may result in deterioration of dispersion on dilution. In many cases, a high modulus and a low critical strain for the linear viscoelastic region implies extensive flocculation and this situation must be avoided in practice.

The critical strain and the storage modulus in the linear region can be used to obtain the cohesive energy of the weakly flocculated structure in suspension concentrates. The cohesive energy, E_c, is related to the stress in the flocculated structure, σ, by the following:

$$E_c = \int_0^{\gamma_c} \sigma \, d\gamma \tag{143}$$

Since $\sigma = G'\gamma$, then

$$E_c = \int_0^{\gamma_c} G'\gamma \, d\gamma$$

$$= \frac{1}{2} G' \gamma_c^2 \tag{144}$$

Depending on the system, an optimum value for E_c is required for maintaining the long-term stability of suspension concentrates. Thus, by carrying out rheological measurements as a function of strain amplitude and frequency, one is able to obtain correlations between the various rheological parameters and the long-term physical stability of the suspension concentrate. These correlations should be established during the formulation of the suspension concentrate and they could be used for further development of the system. In addition, these correlations may be used as guidelines for the formulation of any future suspension concentrate. However, the formulation chemist should be careful in the interpretation of the results since the rheological results depend on several parameters of the system that may simultaneously change during the formulation. The results obtained in a small scale laboratory preparation must be correlated with the properties of the large scale preparation before the optimum rheological parameters are established. This requires more research before it can be used as a quality assurance method. At present dynamic measurements can be used by the formulation chemist to establish the optimum composition of the antisettling system. If such measurements are combined with constant stress results, one may be able to predict the long term physical stability of the suspension concentrate.

7

Microemulsions

I. INTRODUCTION

As mentioned in Chapter 4, one of the most common agrochemical formulations are emulsifiable concentrates (ECs) which when added to water produce oil-in-water (o/w) emulsions either spontaneously or by gentle agitation. These formulations are simple to prepare, whereby the agrochemical is mixed with or dissolved in an oil (usually an aromatic oil such as alkylbenzene) and two emulsifiers are added at sufficient concentration (usually > 5%). When the EC is added to water, the low interfacial tension between oil and water allows spontaneous emulsification to occur. The most commonly used surfactant blends are a mixture of calcium or magnesium alkylarylsulfonate and an ethoxylated nonionic surfactant, such as nonylphenol with several ethylene oxide (EO) units. However, in recent years there was great concern in using ECs in agrochemical formulations for a number of reasons. The use of aromatic oils is undesirable due to their possible phytotoxic effect and their environmental disadvantages.

An alternative and more attractive system to ECs are o/w emulsions, as discussed in Chapter 5. In this case, the pesticide, which may be an oil, is emulsified in water and a water-based concentrated emulsion (EW) is produced. In cases where the pesticide is very viscous or semisolid, a small amount of an oil (which may be aliphatic in nature) may be added before the emulsification process. Unfortunately, EWs suffer from a number of problems such as the difficulty of emulsification and their long-term physical stability. As discussed in Chapter 5, emulsions are thermodynamically unstable systems since their formation involves a high interfacial energy that is not overcome by the relatively small entropy of dispersion (see below). With time, the emulsion tends to reduce its interfacial area by several breakdown processes such as flocculation and coalescence. In addition, most emulsion systems undergo sedimentation or creaming on standing and they require the use of thickeners to modify their rheology. A further complication may arise from Ostwald ripening whereby the smaller droplets dissolve and become deposited on the larger ones. Clearly, to produce a physically stable emulsion of an agrochemical requires a great deal of process control as well prevention of the various breakdown processes.

A very attractive alternative for formulation of agrochemicals is to use microemulsion systems. The latter are single optically isotropic and thermodynamically stable dispersions consisting of oil, water, and amphiphile (one or more surfactants) [191]. As we will see later, the origin of the thermodynamic stability arises from the low interfacial energy of the system which is outweighed by the negative entropy of dispersion. These systems offer a number of advantages over o/w emulsions for the following reasons. Once the composition of the microemulsion is identified, the system is prepared by simply mixing all of the components without the need of any appreciable shear. Due to their thermodynamic stability, these formulations undergo no separation or breakdown on storage (within a certain temperature range depending on the system). The low viscosity of the microemulsion systems ensures their ease of pourability and dispersion on dilution, and they leave little residue in the container. Another main attraction of microemulsions is their possible enhancement of biological efficacy of many agrochemicals. This, as we will see later, is due to the solubilization of the pesticide by the microemulsion droplets.

This chapter summarizes the basic principles involved in the preparation of microemulsions and the origin of their thermodynamic stability. A section will be devoted to emulsifier selection for both o/w and w/o microemulsions. The physical methods that may be applied for characterization of microemulsions will be briefly described. Finally, a section will be devoted to the possible enhancement of biological efficacy using microemulsions. The role of microemulsions in the enhancement of wetting, spreading, and penetration will be discussed. Solubilization is also another factor that may enhance the penetration and uptake of an insoluble agrochemical.

II. BASIC PRINCIPLES OF MICROEMULSION FORMATION AND THERMODYNAMIC STABILITY

As discussed in Chapter 5, consider the process of formation of oil droplets from a bulk oil phase as was schematically represented in Figure 5.1. This process is accompanied by an increase in the interfacial area, ΔA, and hence an interfacial energy, $\Delta A\gamma$. The entropy of dispersion of the droplets is equal to $T\Delta S$ and hence the free energy of formation of the system is given by the expression:

$$\Delta G = \Delta A\gamma - T\Delta S \tag{145}$$

With macroemulsions (EWs), the interfacial energy term is much larger than the entropy term and hence the process of emulsification is nonspontaneous. In other words, energy is needed to produce the emulsion, e.g., by the use of high-speed mixers. Since the free energy of formation of the system is positive, the emulsion tends to break down by flocculation and coalescence, which reduce the interfacial energy. As discussed in Chapter 5, to reduce flocculation and coalescence, one creates an energy barrier between the droplets, thus preventing their close approach. This is achieved by the emulsifiers which could be ionic, nonionic, or polymeric in nature. The electro-

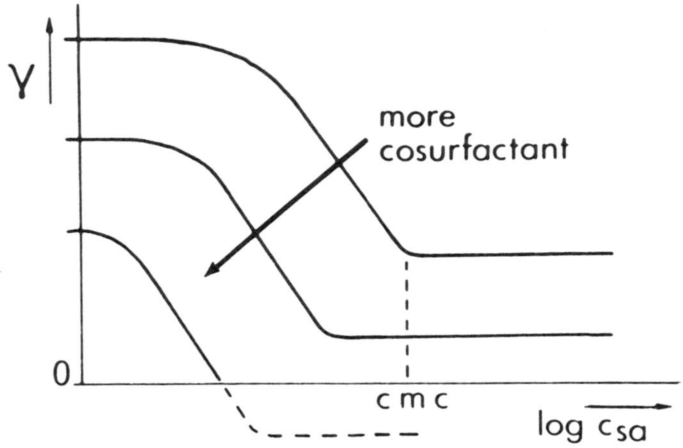

Figure 7.1 $\gamma - \log C$ curves for surfactant plus cosurfactant mixtures.

static or steric repulsion between the droplets prevents their close approach thus preventing flocculation or coalescence, at least within a set period. This explains the limited shelf life of most emulsion systems, which by definition are thermodynamically unstable.

With microemulsions, the interfacial tension is made sufficiently low that the interfacial energy now becomes comparable to or even lower than the entropy of dispersion. In this case, the free energy of formation of the system becomes zero or negative. This explains the thermodynamic stability of microemulsions. Thus the main driving force for microemulsion formation is the ultralow interfacial tension, which is usually achieved by the use of two or more emulsifiers. This can be understood from the effect of surfactant concentration, C, and nature on the interfacial tension, γ, between oil and water. Addition of surfactant to the aqueous or the oil phase causes a gradual lowering of γ, reaching a limiting value at the critical micelle concentration (cmc). Any further increase in C above the cmc causes little or no further decrease in γ. The limiting γ value reached with most single surfactants is seldom lower than 0.1 mN m^{-1}. This value is, in most cases, not sufficiently low for microemulsion formation (which requires γ values of the order of 10^{-2} mN m^{-1} or even lower). However, if a surfactant mixture is used with one component predominantly water soluble, such as sodium dodecyl sulfate, and one predominantly oil-soluble, such as a medium chain alcohol (usually referred to as the cosurfactant), the limiting γ value can reach very low values ($< 10^{-2}$ mNm^{-1}) or even becomes transiently negative [192]. In the latter case, the interface expands spontaneously adsorbing all surfactant molecules till a small positive γ is reached. This behavior is schematically shown in Figure 7.1. As is clearly seen addition of the cosurfactant causes a shift in the $\gamma - \log C$ curve of a surfactant to lower values and the cmc is reduced.

The reason for the reduction in interfacial tension to very low values using two surfactants can be understood from a consideration of the Gibbs adsorption equation [192] that may be extended to a multicomponent system, i.e.,

$$d\gamma = -\sum_i \Gamma_i d\mu_i$$
$$\simeq -\sum_i \Gamma_i RT\, d\ln c_i \tag{146}$$

Integration of Eq. (146) gives

$$\gamma = \gamma_o - \int_0^{c_s} \Gamma_s RT d\ln c_s - \int_0^{c_{co}} \Gamma_{co} RT\, d\ln c_{co} \tag{147}$$

which clearly shows that γ_o is lowered by two terms, both from the surfactant and cosurfactant (which have surface excesses Γ_s and Γ_{co}, respectively). It should be mentioned, however, that the two molecules should become simultaneously adsorbed and should not interact with each other, otherwise they will lower their respective activities. This explains why the two molecules should vary in nature, i.e., one predominantly water soluble and the other oil soluble. It should also be noted that in some cases a single surfactant may be adequate to lower γ sufficiently for microemulsion formation to become possible, e.g., Aerosol OT (sodium diethylhexylsulfosuccinate) and many nonionic surfactants.

The role of the surfactants in microemulsion formation was considered by Schulman and coworkers [193, 194] and later by Prince [195] who introduced the concept of a two-dimensional mixed liquid film as a third phase in equilibrium with both oil and water. This implies that the monolayer of the mixed surfactant film may be represented by a duplex film that gives different properties on the oil and the water side. A two-dimensional surface pressure, π (where π is given by the difference between the interfacial tension of the clean interface and that with the adsorbed surfactant film), describes the property of the film at both sides of the interface. Initially, the flat film will have two different surface pressures at the oil and water sides, namely, π_o' and π_w', respectively. This is due to the different "crowding" of the hydrophobic and hydrophilic components at both interfaces. For example, if the hydrophobic part of the chain is more bulky than the hydrophilic part, crowding will occur at the oil side of the interface and π_o' will be higher than π_w'. This inequality between π_o' and π_w' will result in a stress at the interface that must be relieved by bending. In this case, the film has to be expanded at the oil side of the interface until the surface pressures become equal at both sides of the duplex film, i.e.,

$$\pi_o = \pi_w = 1/2\,(\pi_{\mathrm{o/w}})_a$$

(the subscript a is used to indicate the alcohol cosurfactant, which reduces the interfacial tension on its own right). This leads to the formation of a w/o microemulsion. On the other hand, if $\pi_w' > \pi_o'$, the film has to expand at the water side and an o/w microemulsion is formed. A schematic representation of film bending is shown in Figure 7.2.

According to the mixed film theory, the driving force for film curvature is the stress or the tension gradient, which tends to make

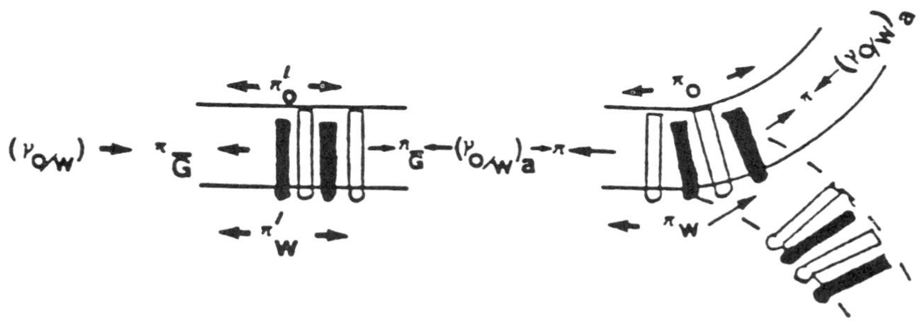

Figure 7.2 Schematic representation of film bending.

the pressure or tension on both sides of the curved film the same. The total interfacial tension, γ_T, is given by the expression:

$$\gamma_T = (\gamma_{\mathrm{o/w}})_a - \pi \tag{148}$$

where π is the two-dimensional spreading pressure of the mixed film. Contributions to π are considered to be the crowding of surfactant and cosurfactant molecules and penetration of the oil phases into the hydrocarbon part of the molecules. According to Eq. (148), if $\pi > (\gamma_{o/w})_a$ then γ_T reaches negative leading to the expansion of the interface until γ_T becomes zero or a small positive value. Since $(\gamma_T)_a$ is of the order of 15–20 mN m^{-1}, surface pressures of that order have to be reached for γ_T to reach an ultralow value that is required for microemulsion formation. This is best achieved by the use of two surfactant molecules as discussed above.

The above simple theory can explain the nature of the microemulsion that is produced when using surfactants with different structures. For example, if the molecules have bulky hydrophobic groups such as Aerosol OT, a w/o microemulsion is produced. On the other hand, if the molecule has bulky hydrophilic chains such as alcohol ethoxylates with high EO units, an o/w microemulsion is produced. These concepts will be rationalized using the packing ratio concept that will be discussed later.

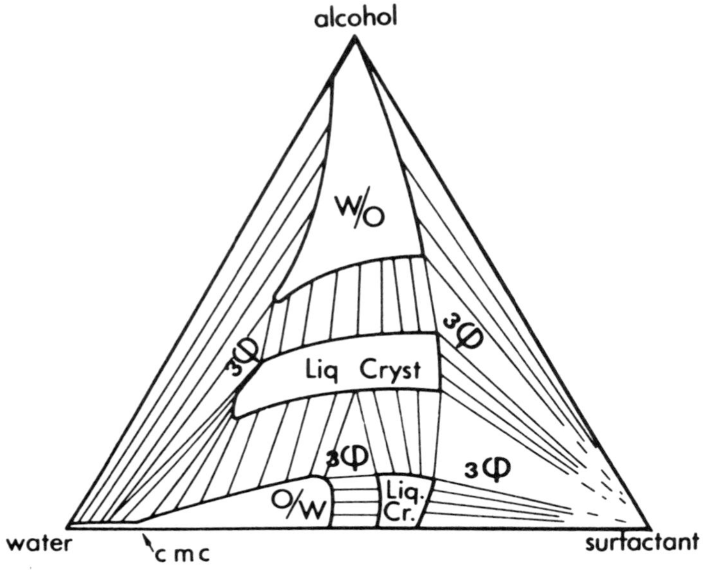

Figure 7.3 Schematic representation of three component phase diagram of water surfactant and cosurfactant (alcohol).

Microemulsions may also be considered as swollen micellar systems as suggested by Shinoda and coworkers [196–198]. These authors considered the phase diagrams of the components of the microemulsion systems. Consider, for example, the phase diagram of a three-component system of water, ionic surfactant (such as sodium dodecyl sulfate), and an alcohol (such as pentanol or hexanol), as shown in Figure 7.3. This phase diagram contains four one phase (1φ) regions: L_1 (o/w normal micelles) at the water corner, L_2 (w/o inverse micelles) at the alcohol corner, and two liquid crystalline regions (a lamellar liquid crystalline region in the middle of the phase diagram and a hexagonal region growing out of the "middle soap" region in the water–surfactant phase diagram) [199]. Between the various one-phase regions, two-phase regions occur where the phases in equilibrium are connected by tie lines [192] and triangular three-phase (3φ) regions.

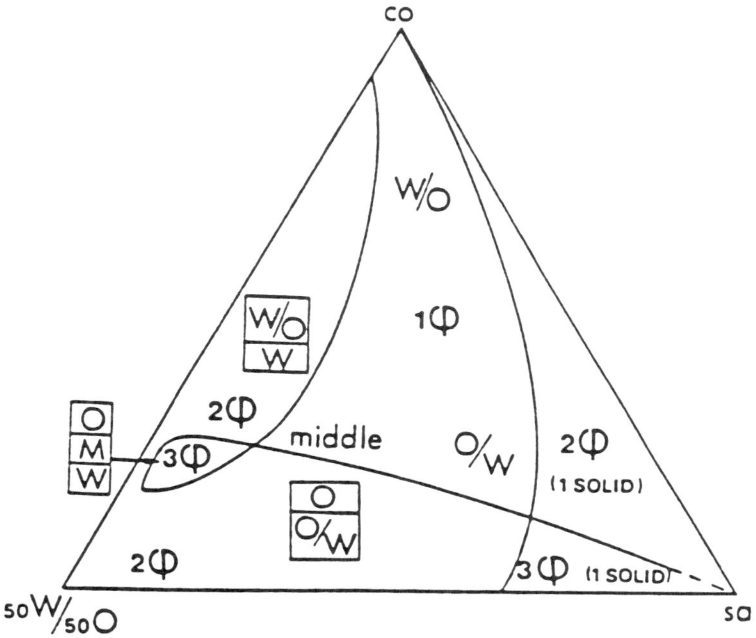

Figure 7.4 Schematic representation of four component phase diagram of oil, water, surfactant, and cosurfactant.

The L_1 (isotropic o/w) and L_2 (isotropic w/o) may be considered as swollen micelles or microemulsions.

Addition of a small amount of oil, miscible with the cosurfactant, changes the phase diagram only slightly. In the presence of substantial amounts of oil, the phase diagram changes significantly as shown in Figure 7.4. The latter figure shows the phase diagram of a four-component system: oil, water, surfactant, and cosurfactant (usually the case with most microemulsions). For simplicity, the phase diagram is displayed in a two-dimensional triangle with the water corner now replaced by a 50:50 water/oil [192]. The o/w microemulsion near the water/surfactant axis is not now in equilibrium with the lamellar phase (as is the case with the three-phase system shown in Figure 7.3), but with a noncolloidal oil + cosurfactant phase. If cosurfactant is added to such a two phase equilibrium at sufficiently high surfactant concentration, all oil is taken up and a one-phase microemulsion appears. However, addition of cosurfactant at low surfactant concentration may lead to separation of an excess aqueous phase before all oil is taken up in the microemulsion. A three-phase system (3φ) is formed containing a microemulsion that cannot be identified as w/o or o/w (bicontinuous or Winsor III phase system). This phase, sometimes referred to as middle-phase microemulsion, is probably similar to the lamellar phase swollen with oil or to a still more irregular intertwining aqueous and oil region (bicontinuous structure). This middle microemulsion phase has a very low interfacial tension with both oil and water (10^{-4}–10^{-2} mN m^{-1}). Further addition of cosurfactant to the three-phase system makes the oil phase disappear and leaves a w/o microemulsion in equilibrium with a dilute aqueous surfactant solution. In the large one-phase region continuous transitions from o/w to middle phase (bicontinuous) to w/o microemulsions are found (Fig. 7.4).

Solubilization and formation of swollen micelles can also be illustrated by considering the phase diagrams of nonionic surfactants containing polyethylene oxide. Such surfactants do not generally need a cosurfactant for microemulsion formation. At low temperatures, the ethoxylated surfactant is soluble in water and at a given concentration is capable of solubilizing a given amount of oil. However, by adding more oil to such solution separation into two phases occurs: o/w solubilized + oil. If the temperature in such a two-phase system is increased, the excess oil may be solubilized. This occurs at the solubilization temperature of the system. Above this temperature an isotropic o/w microemulsion is produced. By further increasing the temperature of this microemulsion, the cloud point of the surfactant is reached and separation into oil + water + surfactant takes place. Thus, an o/w microemulsion is produced between the solubilization temperature and cloud point temperature of the surfactant. This is illustrated in Figure 7.5 which shows the variation of temperature with the weight fraction of the oil (at a constant surfactant concentration). The isotropic o/w microemulsion region is located between the solubilization curve and

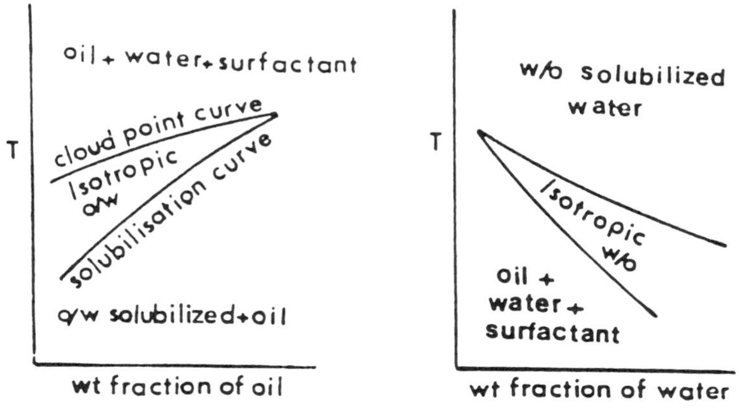

(a) (b)

Figure 7.5 Schematic representation of solubilzation. (a) Oil solubilized in a nonionic surfactant solution. (b) Water solubilized in an oil solution of a nonionic surfactant.

the cloud point curve. This phase diagram shows the temperature range within which a microemulsion is produced. As is clear from Figure 7.5a, this range decreases as the oil weight fraction is increased and above a certain weight fraction (which depends on surfactant concentration) there will be no single isotropic region.

Nonionic ethoxylated surfactants can also be used to produce isotropic w/o microemulsions. A low hydrophobic–lipophilic balance (HLB) number surfactant may be dissolved in an oil and such a solution can solubilize water to a certain extent depending on surfactant concentration. If water is added above the solubilization limit, the system separates into two phases: w/o solubilized + water (see Figure 7.5b). If the temperature of such a two-phase system is reduced an isotropic w/o microemulsion is formed below the solubilization temperature. If the temperature of such system is further reduced below the haze point, separation into water + oil + surfactant occurs. Thus, a w/o microemulsion can be identified between the solubilization and the haze point curves as shown in Figure 7.5b.

It is clear from the above discussion that nonionic surfactants of the ethoxylate type can be used to produce o/w or w/o microemulsions. However, such microemulsions have limited temperature stability and are of limited practical application. Their temperature range of stability may be increased by addition of ionic surfactants, which usually increases the cloud point of the surfactant.

III. FACTORS DETERMINING W/O VS. O/W MICROEMULSION FORMATION

Several factors play a role in determining whether a w/o or an o/w microemulsion is formed. These factors may be considered in the light of the theories described in the previous section. For example, the duplex theory predicts that the nature of the microemulsion formed depends on the relative packing of the hydrophobic and hydrophilic portions of the surfactant molecule which determines the bending of the interface. This can be illustrated by considering an ionic surfactant molecule such as Aerosol OT (diethylhexylsulfosuccinate) which has the following structure.

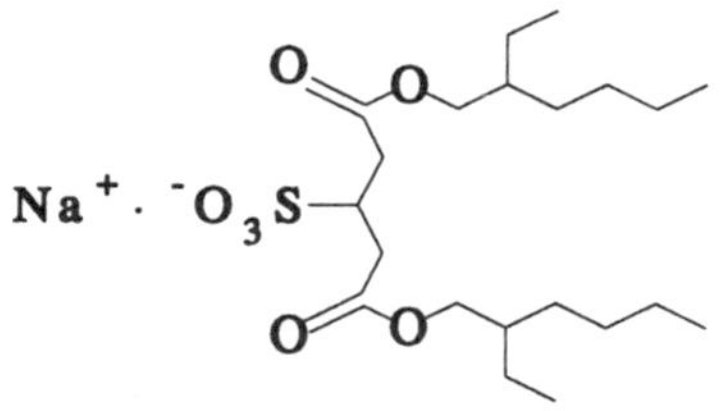

The above molecule has a bulky hydrophobe (two alkyl groups) with a large volume-to-length ratio (*V*/*l*) and a stumpy head group. When this molecule adsorbs at a flat o/w interface, the hydrophobic groups become crowded and the interface tends to bend with the head groups facing inward thus forming a w/o microemulsion. This geometric constraint for the Aerosol OT molecule was considered in detail by Oakenful [200] who showed that the molecule has a *V*/*l* greater than 0.7 which was considered to be necessary for w/o microemulsion formation. For single chain ionic surfactants such as sodium dodecyl sulfate, V/l is less than 0.7 and w/o microemulsion formation requires the presence of a cosurfactant. The latter has the effect of increasing V without affecting l (if the chain length of the cosurfactant does not exceed that of the surfactant). These cosurfactant molecules act as "padding" separating the head groups.

The importance of geometric packing on the nature of microemulsion has been considered in detail by Mitchell and Ninham [201]. According to those authors, the nature of the aggregate unit depends on the packing ratio, P, given by the expression,

$$P = \frac{V}{a_o l_c} \tag{149}$$

where V is the partial molecular volume of the surfactant, a_o is the head group area of a surfactant molecule, and l_c is the maximum chain length. Thus, this packing ratio provides a quantitative measure of the HLB. For values of $P < 1$, normal (i.e., convex) aggregates are predicted, whereas for $P > 1$, inverse drops are expected. The packing ratio is affected by many factors including hydrophobicity of the head group, ionic strength of the solution, pH and temperature, and the addition of lipophilic compounds such as cosurfactants. With Aerosol OT, $P > 1$ since both a_o and l_c are small. Thus this molecule favors the formation of a w/o microemulsion.

The packing ratio also explains the nature of microemulsions formed by using nonionic surfactants. If $V/a_o l_c$ increases with increase of temperature (as a result of the reduction of a_o with temperature), one would expect the solubilization of the hydrocarbon to increase with temperature until $V/a_o l_c$ reaches the value of 1, where phase inversion would be expected. At higher temperatures $V/a_o l_c > 1$ and w/o microemulsion would be expected. Moreover, the solubilization of water would decrease as the temperature rises, as expected.

The influence of surfactant structure on the nature of the microemulsion can be predicted from the thermodynamic theory suggested by Overbeek [202]. According to this theory, the most stable microemulsion would be that in which the phase with the smaller volume fraction, φ, forms the droplets, since the osmotic pressure of the system increases with increasing φ. For a w/o microemulsion prepared using an ionic surfactant, the hard sphere volume is only slightly larger than the water core volume since the hydrocarbon tails may interpenetrate to some extent when two droplets come together. For an o/w microemulsion, on the other hand, the double layer may extend to a considerable extent, depending on the electrolyte concentration. For example, at 10^{-5} mol dm^{-3} 1:1 electrolyte, the double layer thickness is 100 nm. Under these conditions, the effective volume of the microemulsion droplets is much larger than the core oil volume. In 10^{-3} mol dm^{-3}, the double-layer thickness is still significant (10 nm) and the hard sphere radius is increased by 5 nm. Thus, this effect of the double-layer extension limits the maximum volume fraction that can be achieved with o/w microemulsions. This explains why w/o microemulsions with higher volume fractions are generally easier to prepare when compared with o/w microemulsions. Furthermore, establishing a curvature of the adsorbed layer at a given adsorption is easier with water as the disperse phase since the hydrocarbon chains will have more freedom than if they were inside the droplets. Thus to prepare o/w microemulsions at high volume fractions, it is preferable to use nonionic surfactants. As mentioned above, to extend the temperature range, a small proportion of an ionic surfactant must be incorporated and some electrolyte should be added to compress the double layer.

IV. SELECTION OF SURFACTANTS FOR MICROEMULSION FORMULATION

The formulation of microemulsions is still an art, since understanding the interactions, at a molecular level, at the oil and water sides of the interface is far from being achieved. However, some rules may be applied for selection of emulsifiers for formulating o/w and w/o microemulsions. These rules are based on the same principles applied for selection of emulsifiers for macroemulsions described in Chapter 5. Three main methods may be applied for such selection, namely, the hydrophilic–lipophilic balance (HLB), the phase inversion temperature (PIT), and the cohesive energy ratio (CER) concepts. As mentioned earlier, the HLB concept is based on the relative percentage of hydrophilic to lipophilic (hydrophobic) groups in the surfactant molecule. Surfactants with a low HLB number (3–6) normally form w/o emulsions, whereas those with a high HLB number (8–18) form o/w emulsions. Given an oil to be microemulsified, the formulator should first determine its required HLB number. Several procedures may be applied for determining the HLB number depending on the type of surfactant that needs to be used. These procedures have been described in Chapter 5. Once the HLB number of the oil is known one must try to find the chemical type of emulsifier that best matches the oil. Hydrophobic portions of surfactants that are similar to the chemical structure of the oil should be looked at first.

The PIT system provides information on the type of oil, phase volume relationships, and concentration of the emulsifier. The PIT system is established on the proposition that the HLB number of a surfactant changes with temperature and that the inversion of the emulsion type occurs when the hydrophile and lipophile tendencies of the emulsifier balance. At this temperature no emulsion is produced. From a microemulsion viewpoint the PIT has an outstanding feature since it can throw some light on the chemical type of the emulsifier needed to match a given oil. Indeed, the required HLB values for various oils estimated from the PIT system compare very favorably with those prepared using the HLB system described above. This shows a direct correlation between the HLB number and the PIT of the emulsion.

As discussed in Chapter 5, the CER concept provides a more quantitive method for selection of emulsifiers. The same procedure can also be applied for microemulsions.

V. CHARACTERIZATION OF MICROEMULSIONS

Several physical methods may be applied for characterization of microemulsions of which conductivity, light scattering, viscosity, and nuclear magnetic resonance (NMR) are probably the most commonly used. In the early applications of conductivity measurements, the technique was used to determine the nature of the continuous phase. It is expected that o/w microemulsions should give high conductivity, whereas w/o ones should be poorly conducting. Later conductivity measurements were used to give more information on the structure of the microemulsion system, as we will see below. Light scattering, both static (time average) and dynamic (quasi-elastic or photon correlation spectroscopy), is the most widely used technique for measuring the average droplet size and its distribution. This will be discussed in some detail below. Viscosity measurements can be applied to obtain the hydrodynamic radius of microemulsions if the results of viscosity vs. volume fraction could be fitted to some models. The NMR method can be applied to obtain the self-diffusion coefficient of all the components in the microemulsion and this could give information on the structure of the system [203]. Below, the conductivity and light scattering methods will be only described since these are the most practical methods for characterization of agrochemical microemulsions.

A. Conductivity Measurements

One of the most useful application of conductivity was carried out by Shah and Hamlin [204] who followed the change in electrical resistance with the ratio of volume of water to oil (V_w/V_o) for a microemulsion system prepared using the inversion method of Hoar and Schulman [205]. This is illustrated in Figure 7.6 which also indicates the change in optical clarity and birefringence with the ratio of volume of water to oil. At low V_w/V_o, a clear w/o microemulsion is produced that has a high resistance (oil continuous). As V_w/V_o increases, the resistance decreases, and in the turbid or birefringent region hexagonal and lamellar micelles are produced. Above a critical ratio, inversion occurs and the resistance decreases producing an o/w microemulsion.

Conductivity measurements were carried out to study the influence of the nature of the cosurfactant on the structure of w/o microemulsions. This is illustrated in Figure 7.7 which shows the variation of specific conductance on the volume fraction of water, φ_w, for w/o microemulsions prepared using alcohols of various chain lengths [206].

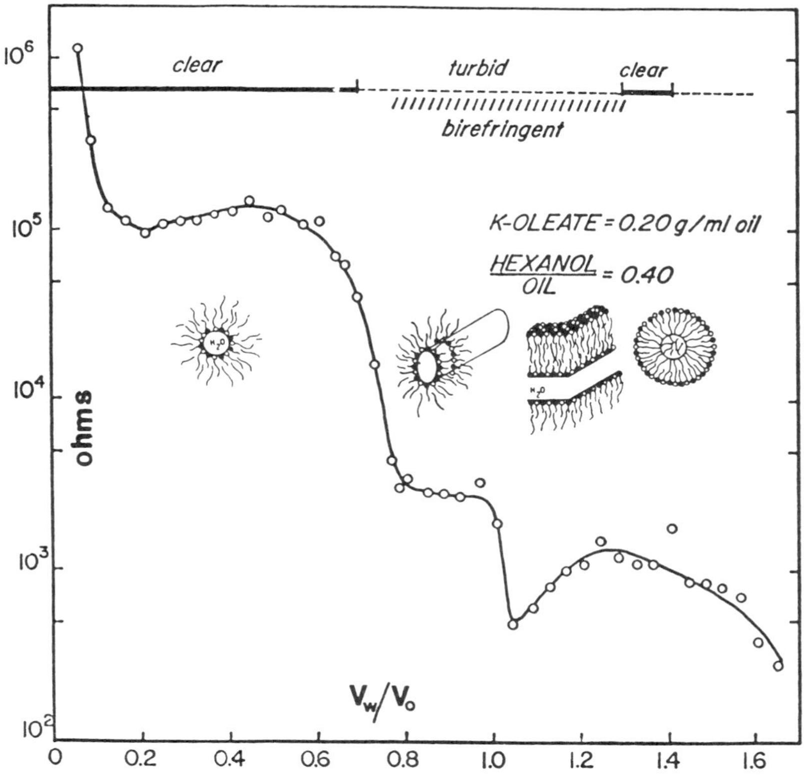

Figure 7.6 Electrical resistance vs. ratio of volume of water to oil in a microemulsion system during inversion.

The cosurfactants can be classified into two main categories. The first category—ethanol, propanol, butanol, and (to some extent) pentanol—show a rapid increase in conductance above a critical φ_w value. These microemulsion systems are described as percolating and they form a bicontinuous structure above the critical water volume fraction. The second category—hexanol and heptanol—shows a low conductance over the whole φ_w range. These systems are "true" microemulsions with definite water cores and are therefore nonpercolating.

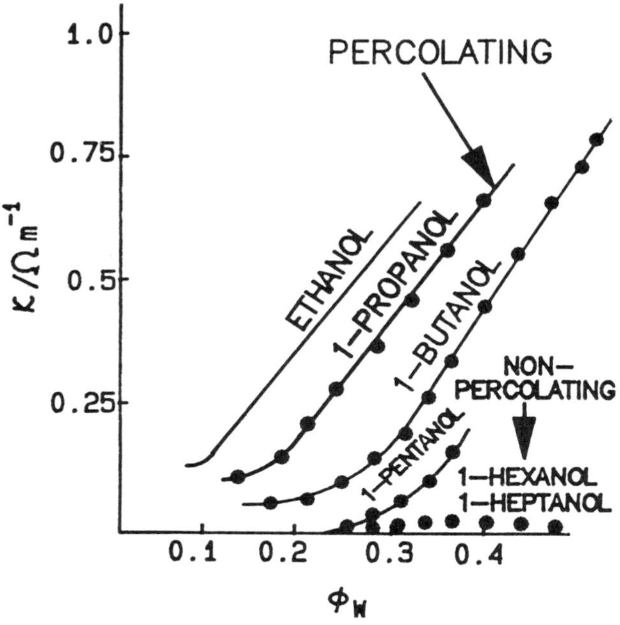

Figure 7.7 Specific conductance vs. volume fraction of water for microemulsions prepared using alcohols of various chain lengths.

B. Light Scattering Measurements

As mentioned above, light scattering measurements provide direct information on the average droplet size. In a dilute noninteracting system of hard spheres, the time average scattering intensity, $I(Q)$, is simply related to the particle size, provided the wavelength of light is more than 10 times the particle diameter, i.e.,

$$I(Q) = [\text{Instrument constant} \times \text{Material constant}]\, NV_p^2 \tag{150}$$

The material constant is determined by the difference in refractive index between the particles and the medium. N is the number of scattering particles with volume V_p. Q is the scattering vector that is determined by the wavelength of light used, λ, and the angle θ at which the measurement is made, i.e.,

$$Q = \frac{4\pi}{\lambda} \sin \frac{\theta}{2} \tag{151}$$

The above method is widely applied for the investigation of many systems such as polymer solutions, micelles, colloidal dispersions, etc. In all these systems, measurements can be made at sufficiently low concentrations to avoid complications produced as a result of particle–particle interactions. The results obtained are then extrapolated to infinite dilution to obtain the desired property such as radius of gyration (for polymers), micelle size (for surfactants), particle radius (for colloidal dispersions), etc. However, the application of this method for microemulsions is not straightforward since dilution of the microemulsion is not possible, as this results in its breakdown. Thus, light scattering measurements have to be made on systems with finite concentrations, where particle–particle interaction is significant and must be taken into account. Thus, Eq. (151) has to be modified to include factors that account for particle interaction. These factors can only be calculated if assumptions are made on the type of interaction. This makes application of light scattering for measurement of droplet size of microemulsions relatively complicated [207].

Another light scattering technique that may be applied for measurement of the hydrodynamic radius, R_h, of microemulsions is photon correlation spectroscopy (PCS). This method was described before in Chapter 6. Basically one measures the intensity fluctuation of scattered light by the droplets as they undergo Brownian motion. From the intensity fluctuation, one may measure the diffusion coefficient of the droplets, D, which is related to R_h, by the Stokes–Einstein equation,

$$D = \frac{kT}{6\pi\eta_o R_h} \tag{152}$$

where k is the Boltzmann constant, T the absolute temperature, and η_o the viscosity of the medium.

Equation (152) can only be applied for a very dilute dispersion. For a more concentrated dispersion, such is the case with micro-emulsions, the collective diffusion coefficient, D, is related to its value at infinite dilution, D_o, through coefficients which depend on the volume fraction φ. For a system that is represented by hard spheres, D and D_o can be related by the following expression (when the interaction between the droplets is repulsive in nature) [208],

$$D = D_o (1 + \alpha\varphi) \tag{153}$$

where α is a constant that equals 1.5 for hard spheres.

VI. ROLE OF MICROEMULSIONS IN ENHANCEMENT OF BIOLOGICAL EFFICACY

The role of microemulsions in enhancement of biological efficiency can be described in terms of the interactions at various interfaces and their effect on transfer and performance of the agrochemical. This will be described in detail in Chapter 8 and only a summary is given here. The application of an agrochemical as a spray involves a number of interfaces, where interaction with the formulation plays a vital role. The first interface during application is that between the spray solution and the atmosphere (air), which governs the droplet spectrum, rate of evaporation, drift, etc. In this respect the rate of adsorption of the surfactant at the air/liquid interface is of vital importance. Since micro-emulsions contain high concentrations of surfactant and mostly more than one surfactant molecule is used for their formulation, then on diluting a microemulsion on application, the surfactant concentration in the spray solution will be sufficiently high to cause efficient lowering of the surface tension γ. As discussed above, two surfactant molecules are more efficient in lowering γ than either of the two components. Thus, the net effect will be production of small spray droplets, which, as we will see later, adhere better to the leaf surface.

In addition, the presence of surfactants in sufficient amounts will ensure that the rate of adsorption (which is the situation under dynamic conditions) is fast enough to ensure coverage of the freshly formed spray by surfactant molecules.

The second interaction is between the spray droplets and the leaf surface, whereby the droplets impinging on the surface undergo a number of processes that determine their adhesion and retention and further spreading on the target surface. The most important parameters that determine these processes are the volume of the droplets, their velocity, and the difference between the surface energy of the droplets in flight, E_o, and their surface energy after impact, E_s. As mentioned above, microemulsions that are effective in lowering the surface tension of the spray solution ensure the formation of small droplets, which do not usually undergo reflection if they are able to reach the leaf surface. Clearly, the droplets need not to be too small otherwise drift may occur. One usually aims at a droplet spectrum in the region of 100–400 μm. As will be discussed in Chapter 8, the adhesion of droplets is governed by the relative magnitude of the kinetic energy of the droplet in flight and its surface energy as it lands on the leaf surface. Since the kinetic energy is proportional to the third power of the radius (at constant droplet velocity), whereas the surface energy is proportional to the second power, one would expect that sufficiently small droplets will always adhere. For a droplet to adhere, the difference in surface energy between free and attached drop ($E_o - E_s$) should exceed the kinetic energy of the drop, otherwise bouncing will occur. Since E_s depends on the contact angle, θ, of the drop on the leaf surface, it is clear that low values of θ are required to ensure adhesion, particularly with large drops that have high velocity. Microemulsions when diluted in the spray solution usually give low contact angles of spray drops on leaf surfaces as a result of lowering the surface tension and their interaction with the leaf surface.

Another factor that can affect biological efficacy of foliar spray application of agrochemicals is the extent to which a liquid wets and covers the foliage surface. This, in turn, governs the final distribution of the agrochemical over the areas to be protected. Several indices may be used to describe the wetting of a surface by the spray liquid, of which the spread factor and spreading coefficient are probably the most useful. The spread factor is simply the ratio between the diameter of the area wetted on the leaf, *D*, and the diameter of the drop, *d*. This ratio is determined by the contact angle of the drop on the leaf surface. The lower the value of θ, the higher the spread factor. As mentioned above, microemulsions usually give low contact angles for the drops produced from the spray. The spreading coefficient is determined by the surface tension of the spray solution as well as the value of θ. Again, with microemulsions diluted in a spray both γ and θ are sufficiently reduced, resulting in a positive spreading coefficient. This ensures rapid spreading of the spray liquid on the leaf surface.

Another important factor for control of biological efficacy is the formation of "deposits" after evaporation of the spray droplets, which ensure the tenacity of the particles or droplets of the agrochemical. This will prevent removal of the agrochemical from the leaf surface by the falling rain. Many microemulsion systems form liquid crystalline structures after evaporation, which have high viscosity (hexagonal or lamellar liquid crystalline phases). These structures will incorporate the agrochemical particles or droplets and ensure their "stickiness" to the leaf surface.

One of the most important roles of microemulsions in enhancing biological efficacy is their effect on penetration of the agrochemical through the leaf. Two effects may be considered that are complementary. The first effect is due to enhanced penetration of the chemical through as a result of the low surface tension. For penetration to occur through fine pores, a very low surface tension is required to overcome the capillary (surface) forces. These forces produce a high pressure gradient that is proportional to the surface tension of the liquid. The lower the surface tension, the lower the pressure gradient and the higher the rate of penetration. The second effect is due to solubilization of the agrochemical within the microemulsion droplet. Solubilization results in an increase in the concentration gradient thus enhancing the flux due to diffusion. This can be understood from a consideration of Fick's first law,

$$J_D = D\left(\frac{\partial C}{\partial x}\right) \tag{154}$$

where J_D is the flux of the solute (amount of solute crossing a unit cross section in unit time), D is the diffusion coefficient and $(\partial C/\partial x)$ is the concentration gradient. The presence of the chemical in a swollen micellar system will lower the diffusion coefficient. However, the presence of the solubilizing agent (the microemulsion droplet) increases the concentration gradient in direct proportionality to the increase in solubility. This is because Fick's law involves the absolute gradient of concentration, which is necessarily small as long as the solubility is small, but not its relative rate. If the saturation is denoted by S, Fick's law may be written as:

$$J_D = D\ 100S\left(\frac{\partial \%S}{\partial x}\right) \tag{155}$$

where $(\partial \%S/\partial x)$ is the gradient in relative value of S. Equation (155) shows that for the same gradient of relative saturation, the flux caused by diffusion is directly proportional to saturation. Hence, solubilization will in general increase transport by diffusion, since it can increase the saturation value by many orders of magnitude (that outweighs any reduction in D.) In addition, the solubilization enhances the rate of dissolution of insoluble compounds and this will have the effect of increasing the availability of the molecules for diffusion through membranes.

8

Role of Surfactants in the Transfer and Performance of Agrochemicals

I. INTRODUCTION

The discovery and development of effective agrochemicals that can be used with maximum efficiency and minimum risk to the user requires the optimization of their transfer to the target during application. In this way the agrochemical can be used in an effective way, thus minimizing any waste during application. Optimization of the transfer of the agrochemical to the target requires careful analysis of the steps involved during application [209]. Most agrochemicals are applied as liquid sprays, particularly for foliar application. The spray volume applied ranges from high values of the order of 1000 liters per hectare (whereby the agrochemical concentrate is diluted with water) to ultralow volumes of the order of 1 liter per hectare (when the agrochemical formulation is applied without dilution). Various spray application techniques are used, of which spraying using hydraulic nozzles is probably the most common. In this case, the agrochemical is applied as spray droplets with a wide spectrum of droplet sizes (usually in the range 100–400 μm in diameter). On application parameters such as droplet size spectrum, their impaction and adhesion, sliding and retention, wetting and spreading are of prime importance in ensuring maximum capture by the target surface as well as adequate coverage of the target surface. These factors will be discussed in some detail below. In addition to these "surface chemical" factors, i.e., the interaction with various interfaces, other parameters that affect biological efficacy are deposit formation, penetration, and interaction with the site of action. As we will see later, deposit formation, i.e., the residue left after evaporation of the spray droplets, has a direct effect on the efficacy of the pesticide, since such residues act as "reservoirs" of the agrochemical and hence control the efficacy of the chemical after application. The penetration of the agrochemical and its interaction with the site of action is very important for systemic compounds. Enhancement of penetration is sometimes crucial to avoid removal of the agrochemical by environmental conditions such as rain and/or wind. All these factors are influenced by surfactants and polymers, as will be discussed in detail below. In addition, some adjuvants that are used in combination with the formulation consist of oils and/or surfactant mixtures. The role of these adjuvants in enhancement of biological efficacy is far from being understood and in most cases they are arrived at by a trial-and-error procedure. A great deal of research is required in this area which would involve understanding the surface chemical processes both static and dynamic, e.g., static and dynamic surface tension and contact angles, as well as their effect on penetration and uptake of the chemical. In recent years, some progress have been made in the techniques that could be applied to such complex problems and these should hopefully lead to a better understanding of the role of adjuvants. The role of these complex mixtures of oils and/or surfactants in controlling agrochemical efficiency is important from a number of points of view. In the first place there is greater demand for reducing the application rate of chemicals and to make better use of the present agrochemicals, e.g., in greater selectivity. In addition, environmental pressure regarding the hazards to the operator and the long-term effects of residues and wastage demands a better understanding of the role of the adjuvants in application of agrochemicals. This should lead to optimization of the efficacy of the chemical as well as reduction of hazards to the operators, the crops, and the environment.

The application of an agrochemical as a spray involves a number of interfaces, where the interaction with the formulation plays a vital role. The first interface during application is that between the spray solution and the atmosphere (air), which governs the droplet spectrum, rate of evaporation, drift, etc. In this respect, the rate of adsorption of the surfactant and/or polymer at the air/liquid interface is of vital importance. This requires dynamic measurements of parameters such as surface tension which will give information on the rate of adsorption. This subject will be dealt with in the first part of this chapter. The second interface is that between the impinging droplets and the leaf surface (with insecticides the interaction with the insect surface may be important). The droplets impinging on the surface undergo a number of processes that determine their adhesion and retention and further spreading on the target surface. The rate of evaporation of the droplet and the concentration gradient of the surfactant across the droplet governs the nature of the deposit formed. These processes of impaction, adhesion, retention, wetting, and spreading will be discussed in subsequent sections of this chapter. The interaction with the leaf surface will be described in terms of the various surface forces involved.

Before describing the above-mentioned processes in detail, it would be useful to say a few words about surfactant molecules and the way they accumulate at various interfaces. This subject has been dealt with in some detail in Chapters 3 and 4. Here only a very brief summary of the main principles will be given for the sake of the analysis that is given in the following sections. As mentioned in Chapter 3, the surfactant molecules accumulate at various interfaces as a result of their dual nature. Basically, a surfactant molecule consist of a hydrophobic chain (usually a hydrogenated or fluorinated alkyl or alkylaryl chain with 8–18 carbon atoms) and a hydrophilic group or chain (ionic or polar nonionic such as polyethylene oxide). At the air/water interface (as for spray droplets) and the solid/liquid interface (such as the leaf surface), the hydrophobic group points toward the hydrophobic surface (air or leaf) leaving the hydrophilic group in bulk solution. This results in lowering of the air/liquid surface tension, γ_{LV}, and the solid/liquid interfacial tension, γ_{SL}. As the surfactant concentration is gradually increased both γ_{LV} and γ_{SL} decrease until the cmc is reached, after which both values remain virtually constant. This situation represents the conditions under equilibrium whereby the rate of adsorption and desorption are the same. The situation under dynamic conditions, such as during spraying may be more complicated since the rate of adsorption is not equal to the rate of formation of droplets. As mentioned in Chapter 2, above the cmc all surfactant molecules form aggregate units or micelles. At concentrations not far from the cmc, these micelles are spherical in nature. At much higher concentrations, cylindrical or rod shaped micelles and lamellar structures are produced. This leads to the formation of liquid crystalline phases with high viscosity, which is the situation under evaporation conditions. As we will see later, these liquid crystalline structures may incorporate the agrochemical (as a solution, particles, or droplets) and they form deposits whose structure and tenacity may play a major role in biological efficacy. Again the presence of micelles and liquid crystalline phases play a major role in the rate of penetration of the chemical and subsequent uptake. This will be discussed in the last section of this chapter.

Another important interaction between the surfactants and the leaf surface is their effect on the wax covering the leaf as well as the possible interactions with the cuticle. This clearly shows that the role of surfactants in controlling the efficacy of the agrochemical can be very crucial and in order to choose the best surfactant for a particular application, one needs to study the above-mentioned interaction in a systematic way.

II. INTERACTIONS AT THE SOLUTION INTERFACE AND THEIR EFFECT ON DROPLET FORMATION

In a spraying process, a liquid is forced through an orifice (the spray nozzle) to form droplets by application of a hydrostatic pressure.

Before describing what happens in a spraying process, it is beneficial to consider the processes that occur when a drop is formed at various time intervals. If the time of formation of a drop is large (greater than, say, 1 min), the volume of the drop depends on the properties of the liquid such as its surface tension and the dimensions of the orifice, but is independent of its time of formation. However, at shorter times of formation of the drop (less than 1 min), the drop volume depends on the time of its formation. Guye and Perrot [210] found that as the speed of formation of the drop is increased its volume increases, passes through a maximum and then decreases. Abbonec [211] derived the following empirical relationship between the weight of a droplet W and its time of formation t,

$$W = a + \frac{b}{t} + \frac{c}{t^2} \tag{156}$$

where a, b, and c are constants, b being proportional to the viscosity of the liquid and c to its density. On differentiation of Eq. (156) with respect to t and equating to zero, the time, t_{max}, at which V reaches a maximum, V_{max}, is given by $2c/b$. Accordingly t_{max} increases with increase of viscosity and decrease of surface tension. The loosening of the drop that occurs when its weight W exceeds the surface force $2\pi r\gamma$ (i.e., $W > 2\pi r\gamma$) progresses at a speed that is determined by the viscosity and surface tension of the liquid. However, during this loosening process the hydrostatic pressure pumps more liquid into the drop and this is represented by a "hump" in the $W - t$ curve. The height of the hump increases with increase of viscosity perhaps because the rate of contraction diminishes as the viscosity rises.

At short t values, W becomes smaller since the liquid in the drop has considerable kinetic energy even before the drop breaks loose. The liquid coming into the drop imparts downward acceleration and this may cause separation before the drop has reached the value given by the equation:

$$W = 2\pi r\,\gamma f\left(\frac{g\rho r^2}{2\gamma}\right) \tag{157}$$

where ρ is the density and r the radius of the orifice. Equation (157) is the familiar equation for calculating the surface or interfacial tension from the drop weight or volume (see Chapter 5).

When the hydrostatic pressure is raised further, i.e., when at even shorter t values than those described above, no separate drops are formed at all and a continuous jet issues from the orifice. Then at even higher hydrostatic pressure, the jet breaks into droplets, which constitutes the phenomenon usually referred to as *spraying*. The process of break-up of jets (or liquid sheets) into droplets is the result of surface forces. The surface area and consequently the surface free energy (area × surface tension) of a sphere is smaller than that of a less symmetrical body. Hence small liquid volumes of other shapes tend to give rise to smaller spheres. For example, a liquid cylinder becomes unstable and divides into two smaller droplets as soon as the length of the liquid cylinder is greater than its circumference [212]. This occurs on accidental contraction of the long liquid cylinder. A prolate spheroid tends to give two spherical drops when the length of the spheroid is greater than three to nine times its width. A very long cylinder with radius r (as, for example, a jet) tends to divide into drops with a volume equal to $(9/2)\pi r^3$. Since the surface area of two unequal drops is smaller than that of two equal drops with the same total volume, the formation of a polydisperse spray is more probable.

The effect of surfactants and/or polymers on the droplet size spectrum of a spray can be, at a first instant, described in terms of their effect on the surface tension. Since surfactants lower the surface tension of water, one would expect that their presence in the spray solution would result in the formation of smaller droplets. This is similar to the process of emulsification described in Chapter 5. As a result of the low surface tension in the presence of surfactants, the total surface energy of the droplets produced on atomization is lower than that in the absence of surfactants. This implies that less mechanical energy is required to form the droplets when a surfactant is present. This leads to smaller droplets at the same energy input. However, the actual situation is not simple since one is dealing with a dynamic situation. In a spraying process a fresh liquid surface is continuously being formed. The surface tension of that liquid depends on the relative ratio between the time taken to form the interface and the rate of adsorption of the surfactant from bulk solution to the air/liquid interface. The rate of adsorption of a surfactant molecule depends on its diffusion coefficient and its concentration (see below). Clearly if the rate of formation of a fresh interface is much faster than the rate of adsorption of the surfactant, the surface tension of the spray liquid will not be far from that of pure water. Alternatively, if the rate of formation of the fresh surface is much slower than that of the rate of adsorption, the surface tension of the spray liquid will be close to that of the equilibrium value of the surface tension. The actual situation is somewhere in between and the rate of formation of a fresh surface is comparable to that of the rate of surfactant adsorption. In this case, the surface tension of the spray liquid will be in between that of a clean surface (pure water) and the equilibrium value of the surface tension which is reached at times larger than that required to produce the jet and the droplets. This shows the importance of measurement of dynamic surface tension and rate of surfactant adsorption.

The rate of surfactant adsorption may be described by application of Fick's first law. When concentration gradients are set up in the system, or when the system is stirred, then the diffusion to the interface may be expressed in terms of Fick's first law [213], i.e.,

$$\frac{d\Gamma}{dt} = \frac{D}{\delta}\frac{N_A}{100} C\,(1 - \theta) \tag{158}$$

where Γ is the surface excess (number of moles of surfactant adsorbed per unit area), t is the time, D is the diffusion coefficient of the surfactant molecule, δ is the thickness of the diffusion layer, N_A is Avogadro's constant, and θ is the fraction of the surface already covered by adsorbed molecules. Equation (158) shows that the rate of surfactant diffusion increases with increase of D and C. The diffusion coefficient of a surfactant molecule is inversely proportional to its molecular weight. This implies that shorter chain surfactant molecules are more effective in reducing the dynamic surface tension. However, the limiting surface tension reached by a surfactant molecule decreases with increase of its chain length and hence a compromise is usually made when selecting a surfactant molecule. Usually one chooses a surfactant with a chain length of the order of 12 carbon atoms. In addition, the higher the surfactant chain length, the lower its cmc (see Chapter 2). Hence lower concentrations are required when using a longer chain surfactant molecule. Again, a problem with longer chain surfactants is their high Krafft temperatures (becoming soluble only at temperatures higher than ambient). Thus, an optimum chain length is usually necessary for optimizing the spray droplet spectrum.

As mentioned above, the faster the rate of adsorption of surfactant molecules, the greater the effect on reducing the droplet size. However, with liquid jets there is an important factor that may enhance surfactant adsorption [214]. Addition of surfactants reduces the surface velocity (which is in general lower than the mean velocity of flow of the jet) below that obtained with pure water. This results from surface tension gradients, which can be explained as follows [214]. Where the velocity profile is relaxing, the surface is expanding, i.e., it is newly formed, and might even approach the composition and surface tension of pure water. A little further downstream, appreciable adsorption of the surfactant will have occurred, giving rise to a back-spreading tendency from this part of the surface in the direction back toward the cleaner surface immediately adjacent to the nozzle. This phenomenon is thus a form of the Marangoni effect (see Chapter 5), which reduces the surface velocity near the nozzle and induces some liquid circulation which accelerates the adsorption of the surfactant molecules by as much as 10 times. This effect casts doubt on the use of liquid jets to obtain the rate of adsorption. Indeed, under conditions of jet formation, it is likely that the surface tension approaches its equilibrium value very closely. Thus, one should be careful in using dynamic surface tension values, as, for example, measured using the maximum bubble pressure method [215].

The influence of polymeric surfactants on the droplet size spectrum of spray liquids is more complicated since adsorbed polymers at the air/liquid interface produce effects other than simply reducing the surface tension. In addition, polymeric surfactants diffuse very slowly to the interface and it is doubtful if they have appreciable effect on the dynamic surface tension. In most agrochemical formulations, polymers are used in combination with surfactants and this makes the situation more complicated. Depending on the ratio of polymer to surfactant in the formulation various effects may be envisaged. If the concentration of the polymer is appreciably greater than the surfactant and interaction between the two components is strong, the resulting "complex" will behave more like a polymer. On the other hand, if the surfactant concentration is appreciably higher than that of the polymer and interaction between the two molecules is still strong, one may end up with polymer–surfactant complexes as well as free surfactant molecules. The latter will behave as free molecules and reduction in the surface tension may be sufficient even under dynamic conditions. However, the role of the polymer-surfactant complex could be similar to the free polymer molecules. The latter produce a viscoelastic film at the air–water interface and that may modify the droplet spectrum and the adhesion of the droplets to the leaf surface. The situation is far from being understood and fundamental studies are required to evaluate the role of polymer in spray formation and droplet adhesion.

The above discussion is related to the case where a polymeric surfactant is used for the formulation of agrochemicals as discussed in Chapters 5 and 6 on emulsions and suspension concentrates. However, in many agrochemical applications high molecular weight materials such as polyacrylamide or polyethylene oxide are sometimes added to the spray solution to reduce drift. It is well known that incorporation of high molecular weight polymers favors the formation of larger drops. The effect can be reached at very low polymer concentrations when the molecular weight of the polymer is fairly high ($> 10^6$). The most likely explanation of how polymers affect the droplet size spectrum is in terms of their viscoelastic behavior in solution. High molecular weight polymers adopt spatial conformations in bulk solution, depending on their structure and molecular weight [216]. Many flexible polymer molecules adopt a random coil configuration that is characterized by a root mean square radius of gyration, R_G. The latter depends on the molecular weight and the interaction with the solvent. If the polymer is in good solvent conditions, e.g., polyethylene oxide in water, the polymer coil becomes expanded and R_G can reach high values, of the order of several tens of nm. At relatively low polymer concentrations, the polymer coils are separated and the viscosity of the polymer solution increases gradually with increase of its concentration. However, at a critical polymer concentration, to be denoted by C^*, the polymer coils begin to overlap and the solutions show a rapid increase in the viscosity with further increase in the concentration above C^*. This concentration C^* is defined as the onset of the semidilute region. C^* decreases with increase in the molecular weight of the polymer and at very high molecular weights it can be as low as 0.01%. Under this condition of polymer coil overlap, the spray jet opposes deformation and this results in the production of larger drops. This phenomenon is applied successfully to reduce drift. Some polymers also produce conformations that approach a rod-like or double-helix structure. An example of this is xanthan gum, which is used with many emulsions and suspension concentrates to reduce sedimentation. If the concentration of such polymer is appreciable in the formulation, then even after extensive dilution on spraying (usually by 100 to 200-fold), the concentration of the polymer in the spray solution may be sufficient to cause production of larger drops. This effect may be beneficial if drift is a problem. However, it may be undesirable if relatively small droplets are required for adequate adhesion and coverage. Again, the ultimate effect required depends on the application methods and the mode of action of the agrochemical. Fundamental studies of the various effects are required to arrive at the optimum conditions. The effect of the various surfactants and polymers should be studied in spray application during the formulation of the agrochemical. In most cases, the formulation chemist concentrates on producing the best system that produces long-term physical stability (shelf life). It is crucial to investigate the effect of the various formulation variables on the droplet spectrum, their adhesion, retention, and spreading. In addition, simultaneous investigations should be made on the effect of the various surfactants on the penetration and uptake of the agrochemical.

III. IMPACTION AND ADHESION

When a drop of a liquid impinges on a solid surface, e.g., a leaf, one of several states may arise depending on the conditions. The drop may bounce or undergo fragmentation into two or more droplets, which in turn may bounce back and return to the surface with a lower kinetic energy. Alternatively, the drop may adhere to the leaf surface after passing through several stages, where it flattens, retracts, spreads, and finally rests to form a hemispherical cap. In some cases, the droplet may not adhere initially but floats as an individual drop for a fraction of a second or even several seconds, and can either adhere to the surface or leave it again.

The most important parameters that determine which one of the above mentioned stages is reached are the mass (volume) of the droplet, its velocity in flight, and the distance between the spray nozzle and the target surface—the difference between the surface energy of the droplet in flight, E_0, and its surface energy after impact, E_s [217]—displacement of air between the droplet and the leaf.

Droplets in the region of 20–50 μm in diameter do not usually undergo reflection if they are able to reach the leaf surface. Such droplets have a low momentum and can only reach the surface if they travel in the direction of the air stream. On the other hand, large droplets of the order of a few thousand μm in diameter undergo fragmentation. Droplets covering the range 100–400 μm, which is the range produced by most spray nozzles, may be reflected or retained depending on a number of parameters such as the surface tension of the spray solution, surface roughness, and elasticity of the drop surface. A study by Brunskill [218] showed that with drops of 250 μm, 100% adhesion is obtained when the surface tension of the liquid was lowered (using methanol) to 39 N m^{-1}, whereas only 4% adhesion occurred when the surface tension, γ, was 57 N m^{-1}. For any given spray solution (with a given surface tension), a critical droplet diameter exists below which adhesion is high and above which adhesion is low. The critical droplet diameter increases as the surface tension of the spray solution decreases. The viscosity of the spray solution has only a small effect on the adhesion of large drops, but with small droplets adhesion increases with increase of the viscosity. As expected, the percentage of adhered droplets decreases as the angle of incidence of the target surface increases.

A simple theory for bouncing and droplet adhesion has been formulated by Hartley and Brunskill [217] who considered an ideal case where there are no adhesion (short range) forces between the liquid and solid substrate and the liquid has zero viscosity. During impaction, the initially spherical droplet will flatten into an oblately spheroidal shape until the increased area has stored the kinetic energy as increased surface energy. This is often followed by an elastic recoil toward the spherical form and later beyond it with the long axis normal to the surface. During this process, energy will be transformed into upward kinetic energy and the drop may leave the surface in a state of oscillation between the spheroidal forms. This sequence was confirmed using high-speed flash illumination.

When the reflected droplet leaves in an elastically deformed condition, the coefficient of restitution must be less than unity since part of the translational energy is transformed to vibrational energy. Moreover, the distortion of droplets involves loss of energy as heat by operation of viscous forces. The effect of increasing the viscosity of the liquid is rather complex, but at a very high viscosity liquids usually have a form of elasticity operating during deformations of very short duration. Reduction of deformation as a result of increase of viscosity will affect adhesion.

As mentioned above, the adhesion of droplets is governed by the relative magnitude of the kinetic energy of the droplet in flight and its surface energy as it lands on the leaf surface. Since the kinetic energy is proportional to the third power of the radius (at constant droplet velocity) whereas the surface energy is proportional to the second power of the radius, one would expect that sufficiently small droplets will always adhere. However, this is not always the case since smaller droplets fall with smaller velocities. Indeed, the kinetic energy of sufficiently small drops, in the Stokesian range, falling at their terminal velocity, is proportional to the seventh power of the radius.

In the 100 to 400 μm range, it is nearly proportional to the fourth power of the radius.

Consider a droplet of radius r (sufficiently small for gravity to be neglected) falling onto a solid surface and spreading with an advancing contact angle, θ_A, and having a spherical upper surface of radius R. The surface energy of the droplet in flight, E_o, is given by:

$$E_o = 4\pi r^2 \gamma_{LA} \tag{159}$$

where γ_{LA} is the liquid/air surface tension.

The surface energy of the spread drop is given as follows:

$$E_s = A_1\gamma_{LA} + A_2\gamma_{SL} - A_2\gamma_{SA} \tag{160}$$

where A_1 is the area of the spherical air/liquid interface, A_2 is that of the plane circle of contact with the solid surface, γ_{SL} is the solid/liquid interfacial tension and γ_{SA} that of the solid/air interface.

From Young's equation,

$$\gamma_{SA} = \gamma_{SL} + \gamma_{LA} \cos\theta \tag{161}$$

Therefore, the surface energy of the droplet spreading on the leaf surface is given by

$$E_s = \gamma_{LA} (A_1 - A_2 \cos\theta) \tag{162}$$

The volume of a free drop is $(4/3)\pi r^3$, whereas that of the spread drop is

$$\pi R^3 \left[(1 - \cos\theta) + \left(\frac{1}{3}\right)(\cos^3\theta) \right]$$

so that

$$\frac{4}{3}\pi r^3 = \pi R^3\,[(1-\cos\theta) + \frac{1}{3}(\cos^3\theta - 1)] \tag{163}$$

and

$$A_1 = 2\pi R^2\,(1-\cos\theta) \tag{164}$$

$$A_2 = \pi R^2 \sin^2\theta \tag{165}$$

Combining Eqs. (160), (162), and (163) one obtains,

$$\begin{aligned} E_s &= \gamma_{LA}\,\pi\,R^2\,[2(1-\cos\theta) - \sin^2\theta\cos\theta] \\ &= \pi^2\,\gamma_{LA}\left(\frac{4}{3}\right)^{2/3}[2(1-\cos\theta) - \sin^2\theta\cos\theta] \\ &\quad \left[(1-\cos\theta) + \left(\frac{1}{3}\right)(\cos^3\theta - 1)\right]^{-2/3} \end{aligned} \tag{166}$$

For a droplet to adhere the difference in surface energy between free and attached drop should exceed the kinetic energy of the drop, otherwise bouncing will occur. As is clear from Eqs. (159) and (166), $E_o - E_s$ depends on γ_{LA} and R. At a given γ_{LA} and R, the smaller the contact angle θ, the larger $E_o - E_s$ and the greater the adhesion. It is perhaps convenient to calculate the ratio $(E_o - E_s)/E_o$, i.e., the minimum energy barrier between attached and free drops, which is necessary for the kinetic energy to overcome, expressed as a fraction of the surface energy of the free drop. From Eqs. (159) and (166),

$$\begin{aligned} \frac{E_o - E_s}{E_o} &= 1 - 0.39\,[2(1-\cos\theta) - \sin^2\theta\cos\theta] \\ &\quad \left[1 - \cos\theta + \frac{1}{3}(\cos^3\theta - 1)\right]^{-2/3} \end{aligned} \tag{167}$$

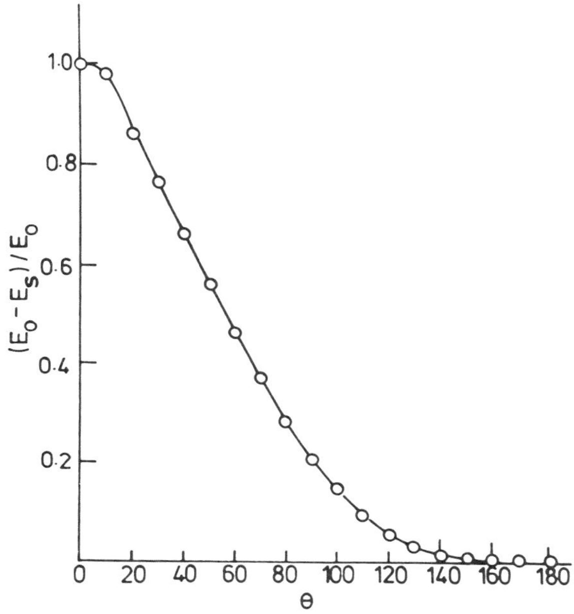

Figure 8.1 Variation of $(E_o - E_s)/E_o$ with the contact angle of a drop on a solid surface.

A plot of $(E_o - E_s)/E_o$ is shown in Figure 8.1, which shows that this ratio decreases rapidly from its value of unity when $\theta = 0$ to a near zero value when $\theta > 160°$.

The above master curve can be used to calculate the critical contact angle required for adhesion of water droplets, with a surface tension $\gamma = 72$ mN m^{-1} at 20°C, of various sizes and velocities. As an illustration, consider a water droplet of 100 μm diameter falling with its terminal velocity v of about 0.25 m s^{-1}. The kinetic energy of the drop is 1.636×10^{-9} J, whereas its surface energy in flight is 2.26×10^{-9} J. The surface energy of the attached drop at which the kinetic energy is just balanced is 2.244×10^{-9} J. The contact angle at which this occurs can be obtained by calculating the fraction $(E_o - E_s)/E_o$ and interpolation using the master curve in Figure 8.1.

Table 8.1 Critical Contact Angle Required for Adhesion of Water Droplets ($\gamma = 72$ mN m^{-1}) at Various Sizes and Velocities

(a) 100 μm droplet $E_o = 2.26 \times 10^{-9}$ J				
Velocity (m s^{-1})	0.25	0.50	0.75	1.00
Kinetic energy	1.64×10^{-11}	6.55×10^{-11}	1.47×10^{-10}	2.62×10^{-10}
$(E_o - E_s)/E_o$	0.0072	0.0298	0.0651	0.1157
θ	160	130	118	105
(b) 200 μm droplet $E_o = 9.05 \times 10^{-9}$ J				
Velocity (m s^{-1})	0.75	1.00	1.25	1.50
Kinetic energy	1.18×10^{-10}	2.09×10^{-9}	3.27×10^{-9}	4.71×10^{-9}
$(E_o - E_s)/E_o$	0.130	0.230	0.361	0.521
θ	102	87	71	54

This gives $(E_o - E_s)/E_o = 0.00723$ and θ about 160°. Thus, providing droplets of this size results in the formation of an angle that is less than 160°, they will stick to the leaf surface. It is thus not surprising that droplets of this size do not need any surfactant for adhesion.

The results of the critical contact angle required for adhesion of water droplets of 100 and 200 μm diameter and various velocities are shown in Table 8.1. It is clear from this table that for any given droplet size the critical contact angle for adhesion decreases with increase in droplet velocity. For example, with the 100 μm droplet, the critical contact angle for adhesion decreases from 160° to 105° as the droplet velocity increases by a factor of four times its terminal velocity. On the other hand, on increase of droplet size the critical contact angle required for adhesion decreases. Thus, for a 200 μm droplet a critical contact angle of 54° is required for a velocity of 1.5 m s^{-1}, which is only twice the terminal velocity. Such velocities are encountered in many spray applications and, therefore, surfactants are required for enhancing adhesion. This is certainly the case for the larger droplets of 300–400 μm, which require much lower contact angles.

It should be mentioned, however, that the above calculations are based on idealized conditions, i.e., droplets falling on a smooth surface. Deviation is expected when dealing with practical surfaces such as leaf surfaces. The latter are rough, containing leaf hairs and wax crystals that are distributed in different ways depending on the nature of the leaf and climatic conditions. Under such conditions, the adhesion of droplets may occur at critical contact angle values that are either smaller or larger than those predicted from the above calculations. The critical θ values will certainly be determined by the topography on the leaf surface. As we will see later, the definition and measurement of the contact angle on a rough surface are not straightforward. In spite of these complications, experimental results on droplet adhesion [219, 220] seem to support the predictions from the above simple theory. These experimental results showed little dependence of adhesion of spray droplets on surfactant concentration. Since with most spray systems the contact angles obtained were lower than the critical value for adhesion (except for droplets larger the 400 μm), then in most circumstances surfactant addition had only a marginal effect on droplet adhesion. However, one should not forget that the surfactant in the spray solution determines the droplet size spectrum. Also addition of surfactants will certainly affect the adhesion of droplets moving at high velocities and on various plant species. The situation is further complicated by the dynamics of the process, which depends on the nature and concentration of the surfactant added. For fundamental investigations, measurements of the dynamic surface tension and contact angle are required on both model and practical surfaces. These measurements are now easy to perform due to technical advances, such as the maximum bubble pressure method for measurement of dynamic surface tension and high speed video equipment for measurement of the dynamic contact angle. Such techniques will enable the formulation chemist and the biologist to understand the role and the function of the surfactant in spray solutions.

IV. DROPLET SLIDING AND SPRAY RETENTION

Many agrochemical applications involve high volume sprays, whereby with continuous spraying the volume of the drops continue to grow in size by impaction of more spray droplets on them and by coal-

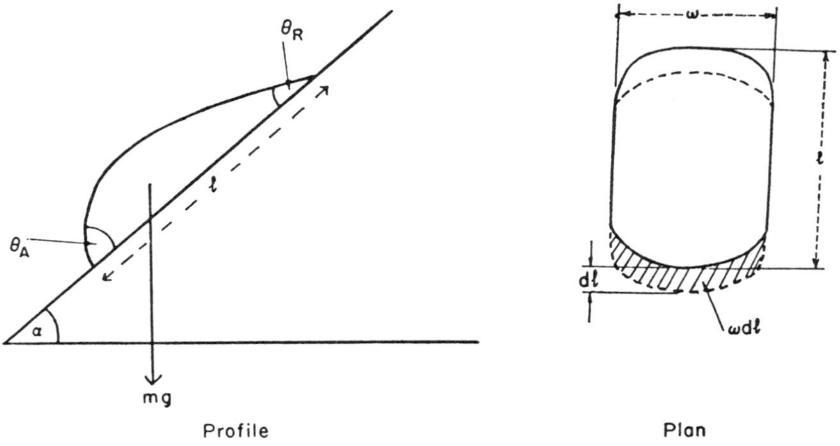

Figure 8.2 Profile and plan view of a drop during sliding.

escence with neighboring drops on the surface. During this process, the amount of spray retained steadily increases provided the liquid drops that are impacted are also retained. However, on further spraying the drops continue to grow in size until they reach a critical value above which they begin to slide down the surface and "drop off," the so called runoff condition. At the point of incipient runoff the volume of the spray retained is a maximum. The retention at this point is governed by the movement of the liquid drops on the solid surface. Bikerman [221] stated that the percentage of droplets sticking to a plant after having touched it should depend on the tilt of the leaf, the size of the droplets, and the contact angle at the plant leaf/droplet/air interface. However, such a process is complicated and governed by many other factors [222] such as droplet spectrum, velocity of impacting droplets, volatility and viscosity of the spray liquid, and ambient conditions.

Several authors [223–226] have tried to relate the resistance to movement of liquid drops on a tilted surface to the surface tension and the contact angles (advancing and receding) of liquid droplets with the solid surface. A detailed analysis was given by Furmidge [222] as summarized below.

Consider a droplet of mass m on a plane surface that is inclined at an angle α form the horizontal (Fig. 8.2). Due to gravity the droplet will start to slide down with a slow constant velocity. Assuming the droplet will have a rectangular plan view (Fig. 8.2), with width w and has moved a distance dl, then the work done by the droplet in moving such a distance, W_g, is given by:

$$W_g = mg \sin \alpha \tag{168}$$

The above force is opposed by the surface force resulting from wetting and dewetting of the leaf surface as the droplet slides downward. In moving down, an area ωdl of the leaf is wetted by the droplet and a similar area is dewetted by the trailing edge. The work of wetting per unit area of the surface is equal to γ_{LA} (cos θ_A + 1), whereas that of dewetting is given by γ_{LA} (cos θ_R + 1), where θ_A and θ_R are the advancing and receding contact angles, respectively. Thus the surface force, W_s, is given by:

$$W_s = \gamma_{LA} \omega dl (\cos \theta_R - \theta_A) \tag{169}$$

At equilibrium,

$$W_g = W_A$$

and,

$$\frac{mg \sin \alpha}{\omega} = \gamma_{LA} (\cos \theta_R - \cos \theta_A) \tag{170}$$

If the impaction of the spray is uniform and the spray droplets are reasonably homogeneous in size, the total volume of spray retained in an area L^2 of surface is proportional to the time of spraying until the time when the first droplet runs off the surface. Also, the volume

V of spray retained per unit area, R, at the moment of incipient runoff is given by:

$$R = \frac{kV}{\omega} \tag{171}$$

where k is a constant. Equation (171) gives the critical relationship of m/ω for the movement of liquid droplets on a solid surface. As the surface is sprayed the adhering drops grow in size until the critical value of m/ω is reached. During this period they remain more or less circular in plan form. Since the droplets are small, the deforming effect of gravity may be ignored and they may be regarded as spherical caps [227] whose volume V is given by the expression,

$$V = \frac{\pi\,(1-\cos\theta)^2\,(2+\cos\theta)\,\omega^3}{24\sin^3\theta}$$

$$= \frac{m}{\rho} \tag{172}$$

Combining Eqs. (170) and (172), it is possible to obtain an expression for ω, the diameter of adhering droplet in terms of the surface forces given above, i.e., γ_{LA} and θ:

$$\omega = \left[\frac{24\sin^3\theta_A\,\gamma_{LA}\,(\cos\theta_R - \theta_A)}{\pi\rho\,(1-\cos\theta_A)^2\,(2+\cos\theta)\,g\sin\alpha}\right]^{1/2} \tag{173}$$

Combining Eqs. (171), (172), and (173), one obtains an expression of spray retention, R, in terms of γ_{LA} and θ, i.e.,

$$R = k\left[\frac{\pi\gamma_{LA}\,(\cos\theta_R - \cos\theta_A)}{24\rho g\sin\alpha}\right]^{1/2}\left[\frac{(1-\cos\theta_A)^2\,(2+\cos\theta_A}{\sin^3\theta_A}\right]^{1/2} \tag{174}$$

The value of k depends on the droplet spectrum, since it relates to the rate of buildup of critical droplets and their distribution. However, Eq. (174) does not take into account the flattening effect of the droplet on impact, which results in reduction of θ and increase of ω above the value predicted by Eq. (173). Thus, Eq. (174) is only likely to be valid under conditions of small impaction velocity. In this case, retention is governed by the surface tension of the spray liquid, the difference between θ_A and θ_R (i.e., the contact angle hysteresis) and the value of θ_A.

Equation (174) can be further simplified by removing the constant terms and standardizing sin α as equal to 1. Further simplification is to replace the second term between square brackets on the right hand side of Eq. (174) by θ_M, the arithmetic mean of θ_A and θ_R. In this way a retention factor, F, may be defined by the following simple expression:

$$F = \theta_M \left[\frac{\gamma_{LA} (\cos \theta_R - \cos \theta_A)}{\rho} \right]^{1/2} \tag{175}$$

Equation (175) shows that F depends on γ_{LA}, the difference between θ_R and θ_A and θ_M. Figure 8.3 shows the variation of F with $(\theta_A - \theta_R)$ at various values of θ_A and choosing some reasonable values of γ_{LA}. As is clear from the curves shown in Figure 8.3, at any given value of θ_A and γ_{LA}, F increases rapidly with increase in $(\theta_A - \theta_R)$ reaches a maximum and then decreases. At any given $(\theta_A - \theta_R)$ and γ_{LA}, F increases rapidly with increase in θ_A (and also θ_M). With systems having the same contact angles, F increases with increase of γ_{LA} but the effect is not very large since $F \propto \gamma_{LA}^{1/2}$. Obviously, any variation in γ_{LA} is accompanied by a change in contact angles and hence one cannot investigate these parameters in isolation. In general, by increasing the surfactant concentration, γ_{LA}, θ_A, and θ_R are reduced. The relative extent to which these three values are affected depends on surfactant nature, its concentration, and the surface properties of the leaf. This is a very complex problem and predictions are almost impossible.

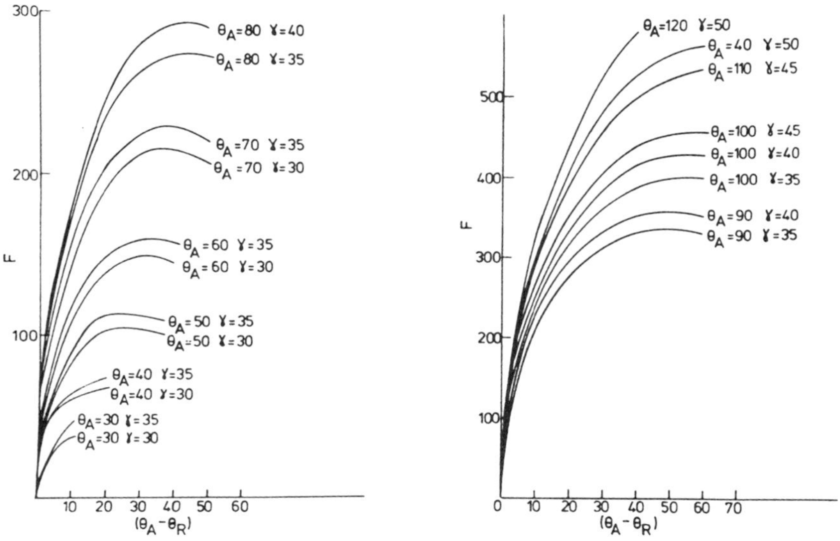

Figure 8.3 Variation of retention factor with contact angle hysteresis.

It should also be mentioned that the above treatment does not take into account the effect of surface roughness and presence of hairs, which play a significant role. A difference in the amount of liquid retained of up to an order of magnitude may be encountered, at constant F value, between, say, a hairy and a smooth leaf. Besides these large variations in surface properties between leaves of various species, there are also variations within the same species depending on age, environmental conditions, and position. However, contact angle measurements on leaf surfaces are not easy and one has to make several measurements and subject the results to statistical analysis. Thus, at best the measured F values can be used as a guideline to compare various surface active agents on leaf surfaces of a particular species that are grown under standard conditions.

Several other factors affect retention, of which droplet size spectrum, droplet velocities, and wind speed are probably the most important. Usually retention increases with reduction of droplet size but is significantly reduced at high droplet velocities and wind speeds. The impact velocity effect becomes more marked as the receding contact angle decreases. Wind reduces the volume of spray that can be retained, particularly when θ_A and θ_R are fairly large because little force is required to remove the drop along the surface. As θ_A and θ_R become small, the wind effect becomes less significant and it becomes negligible when θ_A and θ_R are close to zero. The leaf structure is also important, because less spray is lost due to wind movements from leaves with a very rough surface when compared with smooth leaves. Thus, care should be taken when results are obtained on plants grown under standard conditions, such as glass houses. These results should not be extrapolated to the field conditions because plants grown under normal environmental conditions may have surfaces that are vastly different from those grown in glass houses. In order to obtain a realistic picture on spray retention, measurements should be made on field grown plants and the results obtained may be correlated to those obtained on glass house plants. In this case, it is possible to use glass house plants for selection of surfactants if an allowance is made for the difference between the two sets of results.

V. WETTING AND SPREADING

Another factor that can affect the biological efficacy of foliar spray application of agrochemicals is the extent to which the liquid wets, spreads, and covers the foliage surface. This in turn governs the final distribution of the agrochemical over the area to be protected [227]. The optimum degree of coverage in any spray application depends on the mode of action of the agrochemical and the nature of the pest to be controlled. With nonsystemic agrochemicals, the cover required depends on the mobility or location of the pest. The more static the pest, the greater is the need for complete coverage on those areas of the plant vulnerable to attack. Under those conditions, good spreading of the liquid spray with maximum coverage is required. On the other hand, with systemic agrochemicals satisfactory cover is ensured pro-

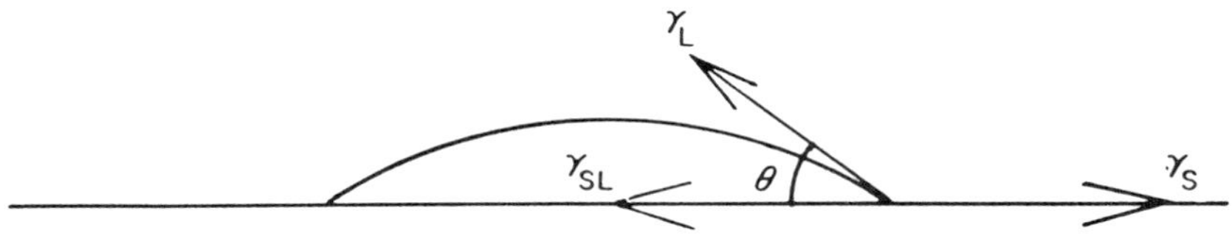

Figure 8.4 Schematic representation of a drop on a leaf surface.

vided the spray liquid is brought into contact with those areas of the plant through which the agrochemical is absorbed. Since, as we will see later, high penetration requires high concentration gradients, an optimum situation may be required here whereby one achieves adequate coverage of those areas where penetration occurs, without too much spreading over the total leaf surface since this usually results in "thin" deposits. These thin deposits do not give adequate "reservoirs," which are sometimes essential to maintain a high concentration gradient, thus enhancing penetration. In addition, thick deposits that are produced from droplets with limited spreading can increase the tenacity of the agrochemical and ensure the longer term protection by the agrochemical. This situation may be required with many systemic fungicides.

Many leaf surfaces represent the most unwettable of most known surfaces. This is due to the predominantly hydrophobic nature of the leaf surface which is usually covered with crystalline wax of straight chain paraffinic alcohols in the range of 24–35 carbon atoms. The crystals may be less than 1 μm thick and only a few μms apart, giving the surface "microroughness," and the "real" area of the surface can be several times the "gross" (apparent) area. When a water drop is placed on a leaf surface, it takes the form of a spherical cap that is characterized by the contact angle θ. This is illustrated in Figure 8.4 for a drop that produces a contact angle < 90°. From the balance of tensions, one obtains the following expression:

$$\gamma_{SA} = \gamma_{SL} + \gamma_{LA} \cos \theta \qquad (176)$$

Equation (176) is the familiar Young's equation, which applies to a liquid drop on a smooth surface.

Wetting is sometimes simply assessed by the value of the contact angle; the smaller the angle, the better the liquid is said to wet the solid. Complete wetting implies a contact angle of zero, whereas complete nonwetting dictates an angle of contact of 180°. However, contact angle measurements are not easy on real surfaces [228] since a great variation in the value is obtained at various locations of the surface. In addition, it is very difficult to obtain an equilibrium value. This is due to the heterogeneity of the surface and its roughness. Thus, in most practical systems such as spray drops on leaf surfaces, the contact angle exhibits hysteresis, i.e., its value depends on the history of the system and varies according to whether the given liquid is tending to advance across or recede from the leaf surface. The limiting angles achieved just prior to movement of the wetting line (or just after movement ceases) are known as the advancing and receding contact angles, θ_A and θ_R, respectively. For a given system, $\theta_A > \theta_R$ and θ can usually take any value between these two limits without discernible movement of the wetting line. Since smaller angles imply better wetting, it is clear that the contact angle always changes in such a direction as to oppose wetting line movement.

The use of contact angle measurements to assess wetting depends on equilibrium thermodynamic arguments [228], which unfortunately is not the real situation. In the practical situation of spraying, the liquid has to displace the air or another fluid attached to the leaf surface and hence measurement of dynamic contact angles, i.e., those associated with moving wetting lines, is more appropriate. Such measurements require special equipment such as video cameras and image analysis and they should enable one to obtain a more accurate assessment of wetting by the spray liquid.

As mentioned above, the contact angle often undergoes hysteresis so that θ cannot be defined unambiguously by experiment. This hysteresis is accounted for by surface roughness, surface heterogeneity and metastable configurations [228]. Surface roughness can be taken into account by introducing a term r in Young's equation, where r is the ratio of real to apparent surface area, i.e.,

$$r\,\gamma_{SA} = r\,\gamma_{SL} + \gamma_{LA}\cos\theta \qquad (177)$$

Thus, the contact angle on a rough surface is given by the expression:

$$\cos\theta = \frac{r\,(\gamma_{SA} - \gamma_{SL})}{\gamma_{LA}} \tag{178}$$

In other words, the contact angle on a rough surface, θ, is related to that on a smooth surface, θ^0, by the equation:

$$\cos\theta = r\cos\theta^0 \tag{179}$$

Equation (179) shows that surface roughness increases the magnitude of $\cos\theta^0$, whether its value is positive or negative. If $\theta^0 < 90°$, $\cos\theta^0$ is positive and it becomes more positive as a result of roughness, i.e., $\theta < \theta^0$ or roughness in this case enhances wetting. In contrast, if $\theta^0 > 90°$, $\cos\theta^0$ is negative and roughness increases the negative value of $\cos\theta^0$, i.e., roughness results in $\theta > \theta^0$. This means that if $\theta^0 > 90°$ roughness makes the surface even more difficult to wet.

The influence of surface heterogeneity was analyzed by Cassie and Baxter [229] who derived the following equation:

$$\cos\theta = a'\cos\theta' + a''\cos\theta'' \tag{180}$$

where a' and a'' are the fractions of the solid surface with different intrinsic contact angles θ' and θ''.

The possibility of adoption of metastable configurations as a result of surface roughness was suggested by Deryaguin [230]. He considered the wetting line to move in a series of thermodynamically irreversible jumps from one metastable configuration to the next.

Assuming an idealized rough surface consisting of concentric patterns of sinusoidal corrugations, as shown in Figure 8.5a, one may relate the apparent contact angle θ to that on a smooth surface θ^0 by the following simple equation [228]:

$$\theta = \theta^0 + \alpha \tag{181}$$

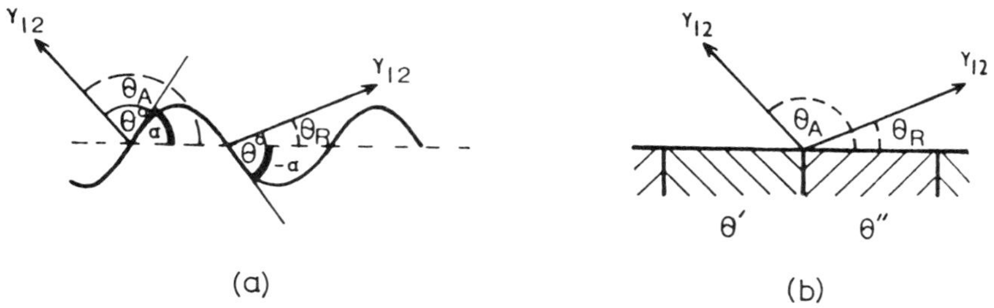

Figure 8.5 Origin of contact angle hysteresis on model surfaces. (a) Rough surface; (b) Heterogeneous surface.

where α is the slope of the solid surface at the wetting line. The value of θ is therefore dependent on the location of the wetting line and, hence, on factors such as drop volume and gravitational forces. A model heterogeneous surface may be represented by a series of concentric bands having alternate characteristic contact angles θ' and θ'', such that $\theta' > \theta''$. This is illustrated in Figure 8.5b. A drop of a liquid placed on this type of a surface will spread or retract until the wetting line assumes some configuration such that $\theta' > \theta > \theta''$.

In spite of the above complications, measurement of contact angles of spray liquids on leaf surfaces is still most useful in defining the wetting and spreading of the spray. A very useful index for measurement of spreading of a liquid on a solid surface is the Harkin's spreading coefficient, S, which is defined by the change in tension when a solid/liquid and liquid/air interfaces are replaced by a solid/air interface [231]. In other words, S is the work required to destroy a unit area each of the solid/liquid and liquid/air interfaces while forming a unit area of the solid/air interface, i.e.,

$$S = \gamma_{SA} - (\gamma_{SL} + \gamma_{LA}) \tag{182}$$

If S is positive, the liquid will usually spread until it completely wets the solid. If S is negative, the liquid will form a nonzero contact angle. This can be clearly shown if Eq. (181) is combined with Young's equation, i.e.,

$$S = \gamma_{LV} (\cos \theta - 1) \tag{183}$$

Clearly, if $\theta > 0$, S is negative and this implies only partial wetting. In the limit $\theta = 0$, S is equal to zero and this represents the onset of complete wetting. A positive S implies rapid spreading of the liquid on the solid surface. Indeed by measuring the contact angle only, one can define a spread factor, SF, which is the ratio between the diameter of the area wetted on the leaf, D, and the diameter of the drop, d, i.e.,

$$SF = \frac{D}{d} \tag{184}$$

Provided θ is not too small (> 5°), the spread factor can be calculated from θ, i.e.,

$$SF = \left[\frac{4 \sin^3 \theta}{(1 - \cos \theta)^2 (2 + \cos \theta)} \right]^{1/3} \tag{185}$$

A plot of SF versus θ is shown in Figure 8.6 which clearly illustrates the rapid increase in SF when $\theta < 35°$. The most practical method of measuring the spread factor is to apply drops of known volume using a microapplicator on the leaf surface. By using a tracer material, such as a fluorescent dye, one may be able to measure the spread area directly using, for example, image analysis. This area can be converted to an equivalent sphere allowing D to be obtained.

An alternative method of defining wetting and spreading is through measurement of the work of adhesion, W_a, which is the work required to separate a unit area of the solid/liquid interface to leave a unit area each of the liquid/air and solid/air, respectively [232], i.e.,

$$W_a = (\gamma_{LA} + \gamma_{SA}) - \gamma_{SL} \tag{186}$$

Again using Young's equation, one obtains the following expression for W_a,

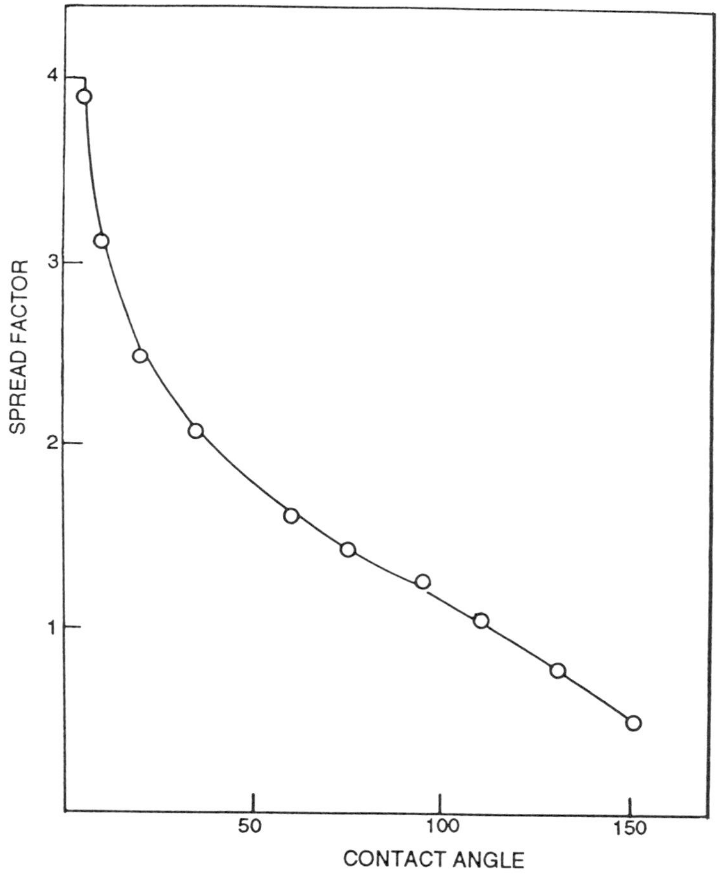

Figure 8.6 Variation of spread factor with contact angle.

$$W_a = \gamma_{LA} (\cos \theta + 1) \tag{187}$$

Another useful concept for assessing the wettability of surfaces is that introduced by Zisman and collaborators [233, 234], namely, the critical surface tension of wetting, γ_c. These authors found that for a given surface and a series of related liquids such as *n*-alkanes, silox-

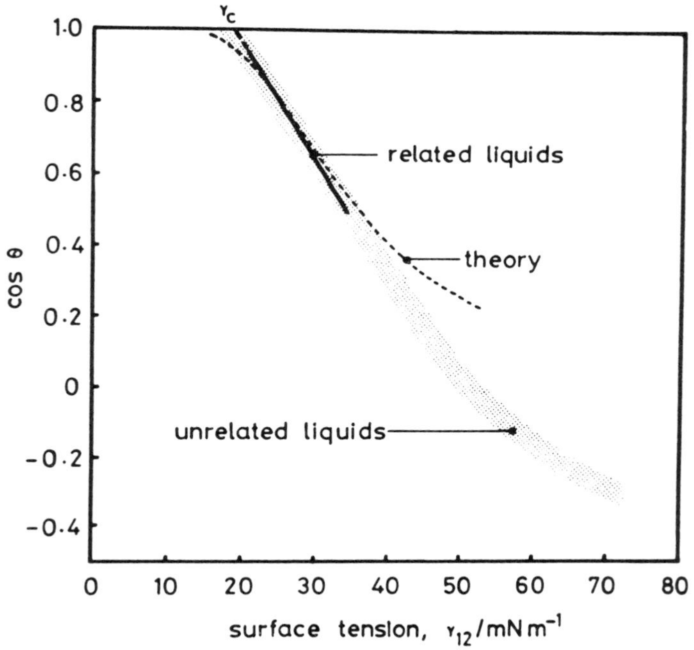

Figure 8.7 Schematic Zisman plot for related liquids (*solid line*) and unrelated liquids of more widely varying surface tension (*shaded band*). The dotted line illustrates the theoretical calculations.

anes, or dialkyl ethers, cos θ is a reasonably linear function of γ_{LA}. This is illustrated in Figure 8.7. for data obtained by Zisman [234] on polytetrafluoroethylene (PTFE). For unrelated liquids of widely ranging surface tension, the line broadens into a band that tends to be curved for high-surface-tension polar liquids. The surface tension at the point where the line cuts the cos θ = 1 axis is known as the critical surface tension of wetting, γ_c. It is the surface tension of a liquid that would just spread to give complete wetting.

Several authors [235, 236] tried to relate the critical surface tension to the solid/liquid interfacial tension, or at least its dispersion component, γ_S^d . Fowkes [236] proposed that the surface and interfacial tensions can be subdivided into independent, additive terms arising from different types of intermolecular interactions. For water, in which both hydrogen bonding (*h*) and dispersion forces (*d*) are active, the surface tension γ is given by,

$$\gamma = \gamma^h + \gamma^d \tag{188}$$

whereas for nonpolar substances such as alkanes and hydrophobic solids, γ is simply equal to γ^d. Thus, for a liquid on a hydrophobic substrate, the solid/liquid interfacial tension may be related to the surface tensions of the solid and liquid by the following:

$$\gamma_{SL} = \gamma_S + \gamma_L - 2\,(\gamma_S^d \gamma_L^d)^{1/2} \tag{189}$$

The third term on the right-hand side of Eq. (189) is simply the geometric mean of the dispersive interactions between liquid and solid. Using Young's equation, one obtains

$$\cos\theta = -1 + \frac{2\,(\gamma_S^d \gamma_L^d)^{1/2}}{\gamma_L} \tag{190}$$

Equation (190) predicts that a plot of cos θ vs. $\gamma_c^{-1/2}$ is linear with intercept $\gamma_c^{-1/2}$ on the cos θ axis. The experimental results seem to support this prediction [235, 236]. A plot of cos θ vs. γ_L using the Fowkes equation is shown schematically by the dotted line in Figure 8.7.

From the above discussion, it is clear that for enhancement of wetting and spreading of liquids on leaf surfaces, one needs to lower the contact angle of the droplets. This is usually achieved by the addition of surfactants, which adsorb at various interfaces and modify the local interfacial tension. The general relationship between the change in contact angle with surfactant concentration and adsorption at the three interfaces (*SA*, *SL*, and *LA*) is obtained by differentiating

Young's equation with respect to ln C (where C is the surfactant concentration) at constant temperature and pressure [237]:

$$\frac{d\gamma_{SA}}{d \ln C} = \frac{d\gamma_{SL}}{d \ln C} + \frac{d\gamma_{LA} \cos \theta}{d \ln C} \tag{191}$$

From the Gibbs adsorption equation (see Chapter 3):

$$\frac{d\gamma}{d \ln C} = -\Gamma RT \tag{192}$$

where Γ is the surface excess, i.e., the number of moles adsorbed per unit area. Equation (192) can be written as (228):

$$\gamma_{LA} \sin \theta \left(\frac{d\theta}{d \ln C} \right) = RT\,(\Gamma_{SA} - \Gamma_{SL} - \Gamma_{LA} \cos \theta) \tag{193}$$

Since γ_{LA} is always positive, $d\theta/d \ln C$ will always have the same sign as the right-hand side of Eq. (193). Three cases may be distinguished. The first case is where

$$\frac{d\theta}{d \ln C} < 0$$

i.e., increasing surfactant concentration enhances wetting; this arises if

$$\Gamma_{SA} < \gamma_{SL} + \gamma_{LA} \cos \theta$$

The second case is where

$$\frac{d\theta}{d \ln C} = 0$$

i.e., increasing surfactant concentration has no effect on wetting. The third case is where

$$\frac{d\theta}{d \ln C} > 0$$

that is,

$$\Gamma_{SA} > \gamma_{SL} + \gamma_{LA} \cos \theta$$

or increasing surfactant concentration reduces wetting. The first and second case are common with predominantly nonpolar low energy surfaces whereas the third case usually occurs with polar substrates. Since most leaf surfaces are nonpolar, low-energy surfaces, increase of surfactant concentration enhances wetting. This explains why most agrochemical formulations contain high concentrations of surfactants to enhance wetting and spreading. However, as we will see later, surfactants play other roles in deposit formation, distribution of the agrochemical on the target surface, and enhancement of penetration of the chemical.

Although the role played by a surfactant is complex, these materials, sometimes referred to as wetting agents or simply adjuvants, need to be carefully selected for optimization of biological efficacy. To date, surfactants are still selected by the formulation chemist on the basis of a trial-and-error procedure. However, some guidelines may be applied in such selection. As discussed in Chapter 5, the HLB system may be initially applied for choosing the most common wetting agents. The latter have HLB numbers between 7 and 9. As discussed in Chapter 2, nonionic surfactants usually have two orders of magnitude lower cmc when compared with their ionic counter parts at the same alkyl chain length. Since the limiting value of the surface tension is reached at concentrations above the cmc, it is clear that many nonionic surfactants are more effective as wetting agents since after dilution of the formulation, the concentration of the nonionic surfactant in the spray solution may be higher than its cmc. However, many nonionic surfactants with HLB numbers in the range 7–9 undergo phase separation at high concentrations and/or temperatures. This may limit their incorporation in the formulation at high concentrations. In some cases addition of a small amount of an ionic surfactant may be beneficial in reducing this phase separation and raising the cloud point of the nonionic surfactant. Thus, many agrochemical formulations contain complex mixtures of surfactants that are carefully arrived at by the formulation chemist. The composition of such mixtures is usually kept confidential.

Another important property of the surfactant that is selected for a given agrochemical is its effect on the leaf structure and the cuticle. Surfactants that cause significant damage to the leaf are described as phytotoxic and in many crops such damage must be avoided. This can sometimes limit the choice, since in certain cases the best wetter may not be the best from the phytotoxity point of view and a compromise has to be made. This shows that selecting the surfactant can be difficult and requires careful investigation of many surface chemical properties as well as its interaction with the leaf surface and the cuticle. In addition, its effect on deposit formation and penetration of the agrochemical needs to be separately investigated.

VI. EVAPORATION OF SPRAY DROPS AND DEPOSIT FORMATION

The object of spraying is often to leave a longlasting deposit of particulate fungicide or insecticide or a residue able to penetrate the cuticle in the case of systemic pesticides and herbicides or to be transferred locally within the crop by its own slower evaporation [238]. The form of residue left by evaporation of the carrier liquid depends to a large extent on the rate of evaporation and most importantly on the nature and concentration of surfactant and other ingredients in the formulations. Evaporation from a spray drop tends to occur most rapidly near the edges since these receive the necessary heat most rapidly from the air by conduction through the dry surround of the leaf. This results in a higher concentration of surfactant at the edge, causing surface tension gradients and convection (arising from the associated density difference). Surface tension gradients cause a Marangoni effect (see Chapter 5) with liquid circulation within the drop that causes the particles to be preferentially deposited at the edge. Convection within the drop leads to preferential precipitation near the edge because the particles can first become "wedged" between the solid/liquid and liquid/air interface.

The type and composition of the spray deposit depends to a large extent on the type of formulation as well as the concentration and type of dispersing agent (for suspensions) or emulsifier (for emulsifiable concentrates and emulsions). Additives such as wetters, humectants, and stickers also affect the nature of the deposit. It should also be mentioned that during evaporation of a spray droplet containing dispersed particles or droplets, these may undergo some physical changes on drying. For example, the solid particles of a suspension may undergo recrystallization forming different shape particles that will affect the final form of the deposit. Both suspension particles and emulsion droplets may also undergo flocculation, coalescence, and Ostwald ripening all of which affect the nature of the deposit. Following such changes during evaporation is not easy and requires special techniques such as microscopy and differential scanning calorimetry.

Another important factor in deposits is the tenacity of the resulting particles or droplets. Strong adhesion between the particles or droplets and the leaf surface is required to prevent removal of these particles or droplets by rain. The adhesion forces between a particle or droplet are determined by the van der Waals attraction as well as the area of contact between the particles and the surface [239]. Several other factors may affect adhesion, i.e., electrostatic attraction, and chemical and hydrogen bonding. The area of contact between the particle and the surface is determined by its size and shape. It is obvious that by reducing the particle size of a suspension, one increases the total area of contact between the particles and the leaf surface, when compared with coarser particles of the same total mass. The shape of the particle also affects the area of contact. For example, flat or cubic-like particles will have larger areas of contact when compared with needle shaped crystals of the same equivalent volume. Several other factors may affect adhesion such as the water solubility of the agrochemical. In general, the lower the solubility, the greater the rain fastness.

One of the most important factors that affect deposit formation is the phase separation that occurs during evaporation. As discussed in Chapter 2, surfactants form liquid crystalline phases when their concentrations exceed a certain value that depends on the nature of the surfactant, its hydrocarbon chain length, and the nature and length of the hydrophilic portion of the molecules. During evaporation, liquid crystals of very high viscosity such as hexagonal or cubic phases may be produced at first. Such highly viscous (and elastic) structures will incorporate any particles or droplets, and act as reservoirs for the chemical. As a result of solubilization of the chemical, penetration and uptake may be enhanced (see below). With further evaporation, the hexagonal and cubic phases may produce lamellar structures with lower viscosity than former phases. Such structures will affect the distribution of particles or droplets in the deposit. Thus, the choice of a particular surfactant for an agrochemical formulation necessitates study of its phase diagram in order to identify the nature of the liquid crystalline phases that are produced on increasing its concentration. The effect of temperature on the liquid crystalline structures is also important. Liquid crystalline structures "melt" above a critical temperature producing liquid phases that contain micelles. These liquid phases have much lower viscosity and hence the particles or droplets of the agrochemical within these liquid phases become mobile. The temperature at which such melting occurs depends on the structure of the surfactant molecule and hence the choice depends on the mode of action of the agrochemical and the environmental conditions encountered such as temperature and humidity. It should also be mentioned that the liquid crystalline structures will be affected by other additives in the formulation such as antifreeze and electrolytes. In addition, the particles or droplets of the agrochemical may affect the liquid crystalline structures produced and this requires a detailed study of the phase diagram in the presence of the various additives as well as in the presence of the agrochemical. Various methods may be applied for such investigations, such as polarizing microscopy, differential scanning calorimetry, and rheology.

VII. SOLUBILIZATION AND ITS EFFECT ON TRANSPORT

Solubilization is usually described as the process of incorporation of an "insoluble substance (usually referred to as the substrate) into surfactant micelles (the solubilizer) [240]. Solubilization may also be referred to as the formation of a thermodynamically stable, isotropic solution of a substance, normally insoluble or slightly soluble in water, by the introduction of an additional amphiphilic component or components. Solubilization can be determined by measuring the concentration of the chemical that can be incorporated in a surfactant solution while remaining isotropic, as a function of its concentration. At concentrations below the cmc, the amount of chemical that can be incorporated in the solution increases slightly above its solubility in water. However, just above the cmc, the concentration of the chemical that can be incorporated in the micellar solution increases rapidly with further increase in surfactant concentration. This rapid increase, just above the cmc, is usually described as the onset of solubilization. One may differentiate three different locations of the substrate in the micelles, as illustrated in Figure 8.8. The most common location is in the hydrocarbon core of the micelle (Fig. 8.8a). This is particularly the case for a lipophilic, nonpolar molecule as is the case with most agrochemicals. Alternatively, the substrate may be incorporated between the surfactant chains of the micelle, i.e., by comicellization (Fig. 8.8b). This is sometimes referred to as penetration in the palisade layer, in which one may distinguish between deep and short penetration. The third way of incorporation is by simple adsorption on the surface of the micelle (Fig. 8.8c). This is particularly the case with polar compounds.

Several factors affect solubilization of which the structure of the surfactant and solubilizate, temperature, and addition of electrolyte are probably the most important [241]. Generalization regarding the manner in which the structural characteristics of the surfactant affect its solubilizing capacity is complicated by the existence of different solubilization sites within the micelles. For deep penetration within the hydrocarbon core of the micelle, solubilization increases with

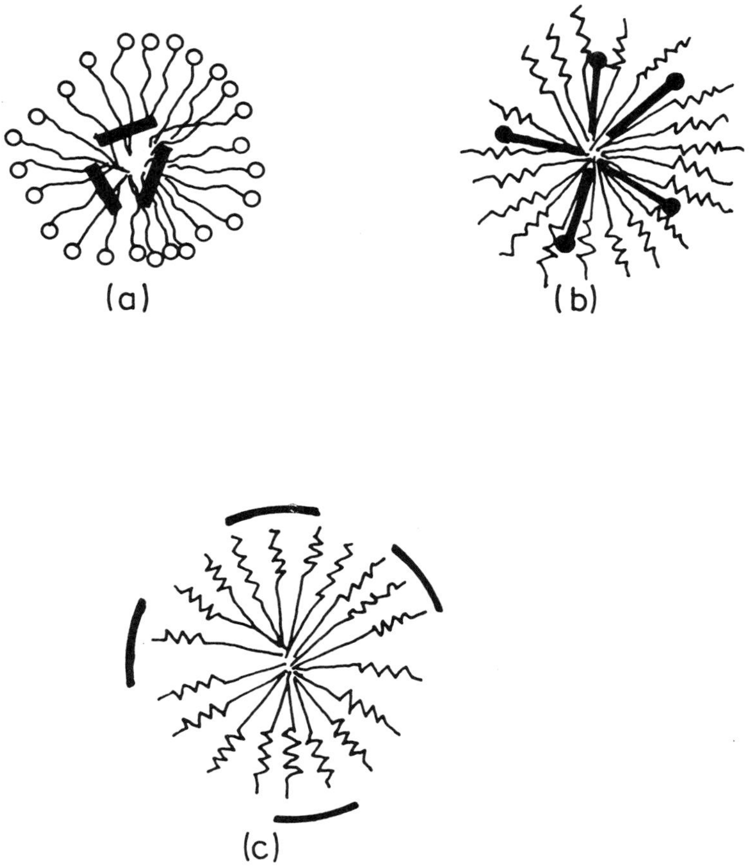

Figure 8.8 Schematic representation of solubilization. (a) In the hydrocarbon core; (b) comicellization; (c) adsorbed on the micelle surface.

increase in the alkyl chain length of the surfactant. On the other hand, if solubilization occurs in the hydrophilic portion of the surfactant molecules, e.g., its polyethylene oxide chain, then the capacity increases with increase in the hydrophilic chain length. The solubilizate structure can also play a major role. For example, polarity and polarizability, chain branching, molecular size and shape, and structure have been shown to have various effects. The temperature also has an effect on the extent of micellar solubilization which is dependent on the structure of the solubilizate and of the surfactant [241]. In most cases, solubilization increases with increase of temperature. This is usually due to the increase of the solubility of the solubilizate and increase of the micellar size with nonionic ethoxylated surfactants. Addition of electrolytes to ionic surfactants usually causes an an increase in the micelle size and reduction in the cmc, and hence an increase in the solubilization capacity. Nonelectrolytes that are capable of being incorporated in the micelle, e.g., alcohols, lead to an increase in the micelle size and hence to an increase in solubilization.

As discussed in Chapter 7, microemulsions, which may be considered as swollen micelles, are more effective in solubilization of many agrochemicals. Oil-in-water microemulsions contain a larger hydrocarbon core than surfactant micelles and hence have larger capacity for solubilizing lipophilic molecules such as agrochemicals. However, with polar compounds o/w microemulsions may not be as effective as micelles of ethoxylated surfactants in solubilizing the chemical. Thus, one has to be careful in applying microemulsions without knowledge of the interaction between the agrochemical and the various components of the microemulsion system.

The presence of micelles or microemulsions (see Chapter 7) will have significant effects on the biological efficacy of an insoluble pesticide. In the first instance, surfactants will affect the rate of solution of the chemical. Below the cmc, surfactant adsorption can aid wetting of the particles and consequently increases the rate of dissolution of the particles or agglomerates [241]. Above the cmc. the rate of dissolution is affected as a result of solubilization. According to the Noyes–Whitney relation [241], the rate of dissolution is directly related to the surface area of the particles A and the saturation solubility, C_s, i.e.,

$$\frac{dC}{dt} = kA\,(C_s - C) \tag{194}$$

where C is the concentration of the solute.

Higuchi [242, 243] assumed that an equilibrium exists between the solute and solution at the solid/solution interface and that the rate of movement of the solute into the bulk is governed by the diffusion of the free and solubilized solute across a stagnant layer. Thus, the effect of surfactant on the dissolution rate will be related to the dependence of that rate on the diffusion coefficient of the diffusing species and not on their solubilities as suggested by Eq. (194). However, experimental results have not confirmed this hypothesis and it was concluded that the effect of solute solubilization involves more steps than a simple effect on the diffusion coefficient. For example, it has been argued that the presence of surfactants may facilitate the transfer of solute molecules from the crystal surface into solution, since the activation energy of this process was found to be lower in the presence of surfactant than its absence in water [244]. On the other hand, Chan et al. [245] considered a multistage process in which surfactant micelles diffuse to the surface of the crystal become adsorbed (as hemimicelles) and form mixed micelles with the solubilizate. The latter is dissolved and it diffuses away into bulk solution, removing the solute from the crystal surface. This multistage process, which directly involves surfactant micelles, will probably enhance the dissolution rate.

Apart from the above effect on dissolution rate, surfactant micelles also affect membrane permeability of the solute [246]. Solubilization can, under certain circumstances, help the transport of an insoluble chemical across a membrane. The driving force for transporting the substance through an aqueous system is always the difference in its chemical potential (or to a first approximation to the difference of its relative saturation) between the starting point and its destination. The principal steps involved are dissolution, diffusion, or convection in bulk liquid and crossing of a membrane. As mentioned above, solubilization will enhance the diffusion rate by affecting transport away from the boundary layer adjacent to the crystal [247, 248]. It should be mentioned, however, that to enhance transport the solution should remain saturated, i.e., excess solid particles must be present since an unsaturated solution has a lower activity.

Diffusion in bulk liquid obeys Fick's first law, i.e.,

$$J_D = D\left(\frac{\partial C}{\partial x}\right) \tag{195}$$

where J_D is the flux of solute (amount of solute crossing a unit cross-section in unit time), D is the diffusion coefficient, and $(\partial C/\partial x)$ is the concentration gradient. The presence of the chemical in a micelle will lower D since the radius of a micelle is obviously greater than that of a single molecule. Since the diffusion coefficient is inversely proportional to the radius of the diffusing particle, D is generally reduced when the molecule is transported by a micelle. Assuming that the volume of the micelle is about 1000 times greater than a single molecule, the radius of the micelle will only be about 10 times larger than that of a single molecule. Thus, D will be reduced by about a factor of 10 when the molecule diffuses within a micelle when compared with that of a free molecule. However, the presence of micelles increases the concentration gradient in direct proportionality to the increase in incorporation of the chemical by the micelle [246]. This is because Fick's law involves the absolute concentration gradient, which is necessarily small as long as the solubility is small, and not its relative rate. If the saturation is represented by S, Fick's law may be written as:

$$J_D = D\ 100\ S\left(\frac{\partial \% S}{\partial x}\right) \tag{196}$$

where $(\partial \% S/\partial x)$ is the gradient in relative value of S. Equation (196) shows that for the same gradient of relative saturation, the flux caused by diffusion is directly proportional to saturation. Hence, solubilization will in general increase transport by diffusion, since it can increase the saturation value by many orders of magnitude (that outweighs the reduction in D).

Solubilization also increases transport by convection since the flux of this process, J_C, is directly proportional to the velocity of the moving liquid and the concentration of the solute, C. Moreover, one would expect that solubilization enhances transport through a membrane [246] by an indirect mechanism. Since solubilization reduces the steps involving diffusion and convection in bulk liquid, it permits application of a greater fraction of the total driving force to transport through the membrane. In this way, solubilization accelerates the transport through the membrane, even if the resistance to this step remains unchanged. It should also be mentioned that enhancement of transport as a result of solubilization does not necessarily involve transport of any micelles. The latter are generally too large to pass through membranes.

The above discussion clearly demonstrates the role of surfactant micelles in the transport of agrochemicals. Since the droplets applied to foliage undergo rapid evaporation, the concentration of the surfactant in the spray deposits can reach very high values that allow considerable solubilization of the agrochemical. This will certainly enhance transport as discussed above. Since the lifetime of a micelle is relatively short, usually less than 1 ms (see Chapter 2), such units break up quickly releasing their contents near the site of action and produce a large flux by increasing the concentration gradient. However, there have been few systematic investigations [249] to study this effect in more detail and this should be a topic of research in the future.

References

1. *McCutcheon's Emulsifiers and Detergents*, McCutcheon Division USA, 1993.
2. Shinoda, K., Nakagawa, T., Tamamushi, B.I., and Isemura, T., *Colloidal Surfactants: Some Physico-chemical Properties*, Academic Press, New York, 1963.
3. Lindman, B., in *Surfactants*, (Tadros, Th.F., ed.), Academic Press, London, 1984.
4. Mukerjee, P., and Mysels, K.J., *Critical Micelle Concentrations of Aqueous Surfactant Systems*, National Bureau of Standards Publication, Washington, D.C., 1971.
5. McBain, J.W., *Trans. Faraday Soc.*, **9**, 99, 1913.
6. Adam, N.K., *J. Phys. Chem.*, **29**, 87, 1925.
7. Hartley, G.S., *Aqueous Solutions of Paraffin Chain Salts*, Hermann and Cie, Paris, 1936.
8. McBain, J.W., *Colloid Science*, Heath, Boston, 1950.
9. Harkins, W.D., Mattoon, W.D., and Corrin, M.L., J. Amer. Chem. Soc., **68**, 220, 1946; *J. Colloid Sci.*, **1**, 105, 1946
10. Debye, P., and Anaker, E.W., *J. Phys. Colloid Chem.*, **55**, 644, 1951.
11. Krafft, F., Ber. *Deutsch Chem. Gessel.*, **32**, 1596 (1899).

12. Shinoda, K., *Principles of Solution and Solubility*, Marcel Dekker, New York, 1974.
13. Clunie, J.S., Goodman, J.F., and Symons, P.C., *Trans. Faraday Soc.*, **65**, 287, 1969.
14. Rosevaar, F.B., *J. Soc. Cosmet. Chem.*, **19**, 581, 1968.
15. Corkill, J.M, and Goodman, J.F., *Adv. Colloid Interface Sci.*, **2**, 297, 1969.
16. Anainsson, E.A.G., and Wall, S.N., *J. Phys. Chem.*, **78**, 1024, 1974; **79**, 857, 1975.
17. Aniansson, E.A.G., Wall, S.N., Almagren, M., Hoffmann, H., Ulbricht, W., Zana, R., Lang, J., and Tondre, C., *J. Phys. Chem.*, **80**, 905, 1976.
18. Rassing, J., Sams, P.J., and Wyn-Jones, E., *J. Chem. Soc., Faraday II*, **70**, 1247, 1974.
19. Jaycock, M.J., and Ottewill, R.H., *Fourth Int. Congress Surface Activity*, **2**, 545, 1964.
20. Okub, T., Kitano, H., Ishiwatari, T., and Isem, N., *Proc. R. Soc.*, **A36**, 81, 1979.
21. Phillips, J.N., *Trans. Faraday Soc.*, **51**, 561, 1955.
22. Kahlweit, M., and Teubner, M., *Adv. Colloid Interface Sci.*, **13**, 1, 1980.
23. Rosen, M.L., *Surfactants and Interfacial Phenomena*, Wiley-Interscience, New York, 1978.
24. Tanford, *The Hydrophobic Effect*, 2nd ed., John Wiley and Sons, New York, 1980.
25. Stainsby, G., and Alexander, A.E., *Trans. Faraday Soc.*, **46**, 587, 1950.
26. Arnow, R.H., and Witten, L., *J. Phys. Chem.*, **64**, 1643, 1960.
27. Guggenheim, E.A., *Thermodynamics*, 5th ed., North Holland, Amsterdam, 1967, p. 45.
28. Gibbs, J.W., *Collected Works*, Vol. 1, Longman, New York, 1928, p. 219.
29. Hough, D.B., and Rendall, H.M., in *Adsorption from Solution at the Solid/Liquid Interface* (Parfitt, G.D., and Rochester, C.H., eds), Academic Press, London, 1983, p. 247.
30. Fuerstenau, D.W., and Healy, T.W., in *Adsorptive Bubble Separation Techniques* (Lemlich, R., ed.), Academic Press, London, 1972, p. 91.
32. Somasundaran, P., and Goddard, E.D., *Mod. Aspects Electrochem.*, **13**, 207, 1979.

33. Tadros, Th.F., *Adv. Chem. Ser.*, **9**, 173, 1975.
34. Robb, D.J.B., and Alexander, A.E., *SCI Monograph*, **25**, 292, 1968.
35. Conner, P., and Ottewill, R.H., *J. Colloid Interface Sci.*, **37**, 642, 1971.
36. Gaudin, A.M., and Fuerstenau, D.W., *Trans. AIME*, **202**, 958, 1955.
37. Clunie, J.S., and Ingram, B.T., in *Adsorption from Solution at the Solid/Liquid Interface*, (Parfitt, G.D., and Rochester, C.H., eds.), Academic Press, London, 1983, p. 105.
38. Tadros, Th.F., in *The Effect of Polymers on Dispersion Properties*, (Tadros, Th.F., ed.), Academic Press, London, 1982.
39. Tadros, Th.F., in *Polymer Colloids*, (Buscall, R., Corner, T., and Stageman, J., eds.), Applied Sciences
40. Jenckel, E., and Rumbach, R., *Z. Electrochem.*, **55**, 612, 1951.
41. Garvey, M.J., Tadros, Th.F., and Vincent, B., *J. Colloid Interface Sci.*, **49**, 57, 1974.
42. van den Boomgaard, Th., King, T.A., Tadros, Th.F., Tang, H.C., and Vincent, B., *J. Colloid Interface Sci.*, **66**, 68, 1978.
43. Tadros, Th.F., and Vincent, B., *J. Colloid Interface Sci.*, **72**, 505, 1979.
44. Fontana, B.J., and Thomas, J.R., *J. Phys. Chem.*, **65**, 480, 1961.
45. Killman, E., Eisenlauer, J., and Korn, M., *J. Polym. Sci.*, **C61**, 413, 1977.
46. Robb, I.D., and Smith, R., Polymer, **18**, 500, 1974; *Eur. Polym. J.*, **10**, 1005, 1974.
47. Barnett, K., Cosgrove, T., Crowley, T.L., Tadros, Th.F., and Vincent, B., in *The Effect of Polymers on Dispersion Properties*, (Tadros, Th.F., ed.), Acdemic Press, London, 1982.
48. Killmann, E., *Polymer*, **17**, 864, 1976.
49. Doroszkowski, A., and Lambourne, R., *J. Colloid Interface Sci.*, **26**, 214, 1968.
50. Garvey, M.J., Tadros, Th.F., and Vincent, B., *J. Colloid Interface Sci.*, **55**, 440, 1976.
51. Pusey, P.N., in *Industrial Polymers, Characterization by Molecular Weight*, (Green, J.H.S., and Dietz, R., eds.), Transcripta Books, London, 1973.
52. Cohen-Stuart, M.A., Waajen, and Dukhin, S.S., *Colloid Polym. Sci.*, **262**, 423, 1984.
53. Cohen-Stuart, M.A., and Mulder, J.W., *Colloids and Surfaces*, **15**, 49, 1985.

54. Jones, H.A., and Fluno, H.J., *J. Econ. Entmol.*, **39**, 735, 1946.
55. Becher, D.Z., in *Encyclopedea of Emulsion Technology*, Vol 2, (Becher, P., ed.), Marcel Dekker, New York, 1985, Chap. 4.
56. Karberg, R.A., and Harris, J.S., U.S. Patent 2,447,475, 1948.
57. Kaberg, R.A., and Harris, J.S., U.S. Patent 2,509,233, 1950
58. Fischer, E., *Farm. Chem.*, **126(3)**, 18, 1963.
59. Lindner, P.L., in *Emulsions and Emulsion Technology*, (Lissant, K.J., ed.), Marcel Dekker, New York, 1974, p. 179.
60. Meusberger, K., in *Advances in Pesticide Formulation Technology*, (Scher, H.B., ed.), ACS Symp. Ser., **254**, 1984.
61. C. Hansen, and A. Beerbower, in *Kirk–Othmer Encyclopedia of Chemical Technology*, 2nd Ed., John Wiley and Sons, New York, 1971.
62. Kaertkemeyer, L., and Amand, J., in *Pesticide Formulation Technology*, SCI Monograph **21**, Gordon and Breach, New York, 1966, p. 28.
63. *Specifications for Pesticides Used in Public Health*, 4th Ed., World Health Organization, Geneva, 1973.
64. Raw, G.R., (ed.), *CIPAC Handbook*, Vol. 1, Heffer, Cambridge, 1970.
65. Behrens, R.W., and Griffin, W.C., *J. Agric. Food Chem.*, **1**, 720, 1953.
66. Lindner, P., in *Herbicides, Fungicides, Formulation Chemistry*, Proc. 2nd Int. IUPAC Congr. Pestic. Chem., Vol. 5, (Tahori, A.S., ed.), Gordon and Breach, New York, 1972, p. 453.
67. Brown, G.L., and Riley, G.C., *Agric. Chem.*, **10(8)**, 34, 1955.
68. Nakagawa, T., in *Nonionic Surfactants*, (Schick, M.J., ed.), Marcel Dekker, New York, 1967, p. 572.
69. Bailey, F.E., and Koleske, J.V., in *Nonioinic Surfactants*, (Schick, M.J., ed.), Marcel Dekker, New York, 1967, p. 794.
70. Carino, L., and Nagy, G., *Pest. Sci.*, **2**, 23, 1971.
71. Gad, J., *Arch. Anat. Physiol.*, 181 (1878).
72. Quinck, G., *Weidemanns Ann.*, **35**, 593 (1888).
73. Lewes, J.B., and Pratt, H.R.C., *Nature*, **171**, 1155, 1953.
74. Garner, F.H., Nutt, C.W., and Mohtadi, M.F., *Nature*, **175**, 603, 1955.
75. Davies, J.T., and Rideal, E.K., *Interfacial Phenomena*, 2nd Ed., Academic Press, New York, 1963.
76. Hartung, H.A., and Rice, O.K., J. *Colloid Interface Sci.*, **10**, 436, 1955.

77. Davis, J.T., and Haydon, D.A., *Proc. Int. Congr. Surface Activity*, 2nd Ed., London, 1957 Vol.1, 1961, p. 417.
78. Kaminski, A., and McBain, J.W., *Proc. R. Soc.*, **A198**, 447, 1949.
79. Ogino, K., and Ota, M., Bull. *Chem. Soc. Jpn.*, **49**, 1187, 1976.
80. Ogino, K., and Umetsu, H., Bull. *Chem. Soc. Jpn.*, **51**, 1543, 1978.
81. Overbeek, J. Th.G., *Faraday Disc. Chem. Soc.*, **65**, 7, 1978.
82. Ruckenstein, E., and Chi, J.C., *J. Chem. Soc. Faraday Trans.*, **71**, 1690, 1975.
83. Lee, G.W.J., and Tadros, Th.F., *Colloids and Surfaces*, **5**, 105, 1982.
84. Lee, G.W.J., and Tadros, Th.F., *Colloids and Surfaces*, **5**, 117, 1982.
85. Lee, G.W.J., and Tadros, Th.F., *Colloids and Surfaces*, **5**, 129, 1982.
86. Boize, L.M., Lee, G.W.J., and Tadros, Th.F., *Pest. Sci.*, **14**, 427, 1983.
87. Cayias, J.L., Schechter, R.S., and Wade, W.H., *ACS Symposium Ser.*, **8**, 234, 1975.
88. Tadros, Th.F., and Vincent, B., in *Encyclopedia of Emulsion Technology*, Vol.1, (Becher, P., ed.), Marcel Dekker, New York, 1983, p. 139.
89. Boffey, B.J., Collison, R., and Lawrence, A.S.C., *Trans. Faraday Soc.*, **55**, 654, 1959.
90. van den Tempel, M., *Rec. Trav. Chim. Pays-Bas*, **72**, 419, 441, 461, 1953.
91. Tadros, Th.F., *SCI Monograph, 28*, 1973, p. 221.
92. Ostwald, W., Z. Phys. Chem., **34**, 493, 1900.
93. Mukerjee, P., *Adv. Colloid Interface Sci.*, **1**, 214, 1967.
94. Mysels, K.J., *Adv. Chem. Ser.*, **86**, 24, 1969.
95. Cockbain, E.C., and McRoberts, J.S., *J. Colloid Sci.*, **8**, 440, 1953.
96. Charles, G.E., and Mason, S.G., *J. Colloid Sci.*, **15**, 236, 1960.
97. Edge, R.M., and Greaves, M., *J. Chem. Eng. Symp.*, **26**, 63, 1967.
98. Hansen, C., and Brown, A.H., *J. Chem. Eng. Symp.*, **26**, 57, 1967.
99. Becher, P., (ed.), *Encyclopedia of Emulsion Technology*, Vol. I, Marcel Dekker, New York, 1983.
100. Walstra, P., in *Encyclopedia of Emulsion Technology*, (Becher, P., ed.), Vol. I, Marcel Dekker, New York, 1983, Chap. 2.
101. van den Tempel, M., *Proc. of the 3rd Conf. Surface Activity*, Vol. 2, Cologne, 1960, p. 573.
102. Prins, A., Acuri, C., and van der Tempel, M., *J. Colloid Interface Sci.*, **24**, 84, 1967.
103. Lucassen, J., in *Physical Chemistry of Surfactant Action*, Vol. 10, (Lucassen-Reynders, E.H., ed.), Marcel Dekker, New York, 1979.

104. Davies, J.T., *Turbulance Phenomenon*, Academic Press, New York, 1972, Chaps. 8 to 10.
105. Chandrosekhav, S., *Hydrodynamics and Hydrodynamic Instability*, Cleeverdon, Oxford, 1961, Chaps. 10 to 12.
106. Lord Rayleigh, *Phil. Mag.*, **14**, 184 (1884).
107. Lord Rayleigh, *Phil. Mag.*, **34**, 145 (1892).
108. Ruscheidt, F. D., and Mason, G., *J. Colloid Interface Sci.* **17**, 260, 1982.
109. Griffin, W. C., *J. Cosmet. Chemists*, **1**, 311, 1949.
110. Griffin, W. C., *J. Cosmet. Chemists*, **5**, 249, 1954.
111. Shinoda, K., *J. Colloid Interface Sci.*, **25**, 396, 1967.
112. Shinoda, K., and Saito, H., *J. Colloid Interface Sci.* **34**, 238, 1969.
113. Shinoda, K., Proc. Int. Congr. Surface Activity, 5th Ed., Vol. 2, Butterworth, London, 1968, p. 295.
114. Beerbower, A., and Hill, M. W., *Am. Cosmet. Pref.*, **87**, **(6)**, 85, 1972.
115. Davies, J. T., *Proc. Int. Congr. Surface Act.*, 2nd Ed., Vol. I., Butterworth, London, 1959, p. 426.
116. Davies, J. T., and Rideal, E. K., *Interfacial Phenomen*, Academic Press, New York, 1961.
117. Becher, P., and Birkmeer, R. L., *J. Amer. Oil Chem. Soc.*, **41**, 169, 1964.
118. Shinoda, K., Saito, H., and Arai, H., *J. Colloid Interface Sci.*, **35**, 624, 1971.
119. Parkinson, C., and Sherman, P., *J. Colloid Interface Sci.* **32**, 642, 1970.
120. Parkinson, C., and Sherman, P., *J. Colloid Interface Sci.* **41**, 328, 1972.
121. Matsumoto, S., and Sherman, P., *J. Colloid Interface Sci.* **33**, 294, 1970.
122. Winsor, P., *Ind. Eng. Chem. Prod. Res. Develop.*, **8**, 2, 1969.
123. Hildebrand, J. H., *Solubility of Non-Electrolytes*, 2nd Ed., Rheinhold, New York, 1936.
124. Hansen, C. M., J. *Paint. Technol.*, **39**, 305, 1967.
125. Barton, A. F. M., *Handbook of Solubility Parameters and other Cohesive Parameters*, CRC Press, New York, 1983.
126. Scheutjens, J. H. M., Fleer, G. J., and Vincent, B., *ACS Symp. Ser.*, **240**, 245, 1984.
127. Hamaker, H. C., *Physica.*, **4**, 1058, 1937.

128. Deryaguin, B. V., and Landau, L., *Acta Physico Chem.*, USSR., **14**, 633, 1949.
129. Verwey, E. J. W., and Overbeek, J. Th. G., *Theory of Stability of Lyophobic Colloids*, Elsevier, Amsterdam, 1948.
130. Napper, D. H., *Polymeric Solubilisation of Dispersion*, Academic Press, London, 1983.
131. Vincent, B., in *Surfactants*, (Tadros, Th. F., ed.), Academic Press, London, 1984.
132. Davis, S. S., and Smith, A., in *Theory and Practice of Emulsion Technology*, (Smith, A. L., ed.), Academic Press, London, 1974, p. 285.
133. Deryaguin, B.V., and Obucher, E., *J. Colloid Chem.*, **1**, 385, 1930; Deryaguin, B.V., and Scherbaker, R. C., *Kolloid Z.*, **23**, 33, 1961.
134. Friberg, S., Jansson, P.O., and Cederberg, E., *J. Colloid Interface Sci.*, **55**, 614, 1976.
135. Wilhelmy, L., *Ann. Phys.*, **119**, 177 (1863).
136. Bashforth, F., and Adams, J.C., *An Attempt to Test the Theories of Capillary Action*, Cambridge University Press, New York, 1883.
137. Niederhauser, D.O., and Bartell, F.E., *Report of Progress, Fundamental Research on Occurence of Petroleum*, Publication of the American Petroleum Institute, Lord Baltimore Press, Baltimore, 1950, p. 114.
138. Du Nouy, P.L., *J. Gen. Physiol.*, **1**, 521, 1919.
139. Harkins, W.D., and Jordan, H.F., *J. Amer. Chem. Soc.*, **52**, 1715, 1930.
140. Freud, B.B., and Freud, H.Z., *J. Amer. Chem. Soc.*, **52**, 1772, 1930.
141. Harkins, W.D., and Brown, F.E., *J. Amer. Chem. Soc.*, **41**, 499, 1919.
142. Lando, J.L., and Oakley, H.T., *J. Colloid Interface Sci.*, **25**, 526, 1967.
143. Wilkinson, M.C., and Kidwell, R.L., *J. Colloid Interface Sci.*, **35**, 114, 1971.
144. Vonnegut, B., *New Sci. Intrum.*, **13**, 6, 1942.
145. Criddle, D.W., *The Viscosity and Elasticity of Interfaces in Rheology, Theory and Applications*, Vol. 3, (Eirich, F.R., ed.), Academic Press, 1960, Chap. 11, p. 429.
146. Edwards, D.A., Brenner, H., and Wasan, D.T., *Interfacial Transport and Rheology*, Butterworths-Heinman, Boston, 1991.
147. Tadros, Th.F., *Adv. Colloid Interface Sci.*, **12**, 141, 1980.
148. Tadros, Th.F., *Adv. Colloid Interface Sci.*, **32**, 205, 1990.
149. Tadros, Th.F., *Pest. Sci.*, **26**, 51, 1989.

150. Tadros, Th.F., in *Solid/Liquid Dispersions*, (Tadros, Th.F., ed.), Academic Press, London, 1967.
151. Parfitt, G.D., *Dispersions of Powders in Liquids*, Applied Science Publishers, 1973.
152. Patton, *Paint Flow and Pigment Dispersion*, Interscience, New York, 1964.
153. Washburn, E.D., *Phys. Rev.*, **17**, 273, 374, 1921; Rideal, E.K., Phil. Mag., **44**, 1152, 1922.
154. Rehbinder, P.A., *Colloid J. USSR*, **20**, 493, 1958.
155. Rehbinder, P.A., and Likhtman, V.I., *Proc. 2nd Int. Conference Surface Activity*, Vol. 3, Butterworth, London, 1957, p. 157.
156. Schukin, E.D., and Rehbinder, P.A., *Colloid J. USSR*, **20**, 601, 1958.
157. Schukin, E.D., *Sov. Sci. Rev.*, **3**, 157, 1972.
158. Bartenev, G.M., Iudena, I.V., and Rebinder, P.A., *Colloid J. USSR*, **20**, 611, 1958.
159. Heath, D., Knott, R.D., Knowles, D.A., and Tadros, Th.F., *ACS Symp. Ser.*, **254**, 11, 1984.
160. Ostwald, W., *Z. Phys. Chem.*, **34**, 493, 1900.
161. Buckley, H.E., *Crystal Growth*, John Wiley and Sons, London, 1951.
162. Khaminski, E.V., *Crystallization from Solution*, Consultants Bureau, New York, 1969.
163. Nancollas, G.H. anmd Purdie, N., *Q. Rev.*, **18**, 1, 1964.
164. Tadros, Th.F., in *Particle Growth in Suspensions*, (Smith, A.L., ed.), SCI Monograph, Academic Press, London, 1973.
165. Maude, A.D., and Whitmore, R.L., *Brit. J. Appl. Phys.*, **9**, 477, 1958.
166. Bachelor, G.K., *J. Fluid Mech.*, **52**, 245, 1972.
167. Reed, C.C., and Anderson, J.L., *AIChE J.*, **26**, 814, 1980.
168. Buscall, R., Goodwin, J.W., Ottewill, R.H., and Tadros, Th.F., *J. Colloid Interface Sci.*, **85**, 78, 1982.
169. Krieger, I.M., *Adv. Colloid Interface Sci.*, **3**, 45, 1971.
170. van Olphen, H., *Clay Colloid Chemistry*, John Wiley and Sons, New York, 1963.
171. Norrish, K., *Disc. Faraday Soc.*, **18**, 120, 1954.
172. Tadros, Th.F., *Colloids and Surfaces*, **18**, 427, 1986.
173. Tadros, Th.F., *Effect of Polymers on Dispersion Properties*, (Tadros, Th.F., ed.), Academic Press, London, 1982.
174. Asakura, S., and Oosawa, F., *J. Chem. Phys.*, **22**, 1255, 1954; *J. Polym. Sci.*, **33**, 183, 1958.

175. Heath, D., Knott, R.D., Knowles, D.A., and Tadros, Th.F., *ACS Symp. Ser.*, **254**, 2, 1984.
176. Hunter, R.J., *Zeta Potential in Colloid Science: Principles and Applications*, Academic Press, London, 1981.
177. Gregg, S.J., and Sing, K.S.W, *Adsorption, Surface Area and Porosity*, Academic Press, London, 1967.
178. Kipling, J.J., and Wilson, R.B., *J. Appl. Chem.*, **10**, 109, 1960.
179. Greenwood, F.G., Parfitt, G.D., Picton, N.H., and Wharton, D.G., *Adv. Chem. Ser.*, **79**, 135, 1968.
180. Clunie, J.S., and Ingram, B.T., *Adsorption from Solution at the Solid/Liquid Interface*, (Parfitt, G.D., and Rochester, C.H., eds.), Academic Press, London, 1983, Chap. 3.
181. Fleer, G.J., Cohen-Stuart, M.A., Scheutjens, J.M.H.M., Cosgrove, T., and Vincent, B., *Polymers at Interfaces*, Chapman and Hall, London, 1993.
182. Cosgrove, T., Crowley, T.L., Vincent, B., Barnett, K.G., and Tadros, Th.F., *Faraday Symp. Chem. Soc.*, **16**, 101, 1981
183. Cohen-Stuart, M.A., Scheutjens, J.M.H.M, and Fleer, G.J., *J. Polym. Sci., Polym. Phys. Ed.*, **18**, 559, 1980.
184. Robb, I.D., *Comprehensive Polym. Sci.*, **2**, 733, 1989.
185. Cosgrove, T., and Rayen, K., *Lamgmuir*, **6**, 136, 1990.
186. von Smoluchowski, M., *Physik. Z.*, **17**, 557, 585, 1916.
187. Whorlow, R.W., *Rheological Techniques*, Ellis Horwood, Chichester, 1980.
188. Casson, N. *Rheology of Disperse Systems*, (Mill, C.C., ed.), Pergamon Press, Oxford, 1959, p. 84.
189. Heath, D., Thomas, P.K., Warrington, R., and Tadros, Th.F., *ACS Symp. Ser.*, **254**, 11, 1984.
190. Ferry, J.D., *Viscoelastic Properties of Polymers*, John Wiley and Sons, New York, 1980.
191. Danielsson, I., and Lindman, B., *Colloids and Surfaces*, **3**, 391, 1981.
192. Overbeek, J.Th.G., de Bruyn, P.L., and Verhoecks, F., in *Surfactants*, (Tadros, Th.F., ed.), Academic Press, London, 1984, p. 111.
193. Bowcott, J.E., and Schulman, J.H., *J. Electrochem.*, **54**, 283, 1955.
194. Sculman, J.H., Stockenius, W., and Prince, L.M., *J. Phys. Chem.*, **63**, 1677, 1959.
195. Prince, L.M, *J. Colloid Interface Sci.*, **23**, 165, 1967.
196. Shinoda, K., and Friberg, S., *Adv. Colloid Interface Sci.*, **4**, 281, 1975.

197. Saito, H., and Shinoda, K., *J. Colloid Interface Sci.*, **24**, 10, 1967; **26**, 70, 1968.
198. Shinoda, K., and Kunieda, H., *J. Colloid Interface Sci.*, **42**, 381, 1973.
199. Danielsson, I., Friman, R., and Sjoblom, J., *J. Colloid Interface Sci.*, **85**, 42, 1982.
200. Oakenfull, D., *J. Chem. Soc. Faraday Trans. I*, **76**, 1875, 1980.
201. Mitchell, D.J., and Ninham. B.W., *J. Chem. Soc. Faraday Trans. II*, **77**, 601, 1981.
202. Overbeek, J.Th.G., *Faraday Disc. Chem. Soc.*, **65**, 7, 1978.
203. Lindman, B., Stilbs, P., and Moseley, M.E., *J. Colloid Interface Sci.*, **83**, 569, 1981.
204. Shah, D.O., and Hamlin, R.M., *Science*, **171**, 483, 1971.
205. Hoar, T.P., and Schulman, J.H., *Nature*, **152**, 102, 1943.
206. Clausse, M., Peyerlesse, J., Boned, C., Heil, J., Nicolas-Margantine, L., and Zrabda, A., in *Solution Properties of Surfactants*, Vol. 3, (Mittal, K.L., and Lindman, B., eds.), Plenum Press, 1984, p. 1583.
207. Baker, R.C., Florence, A.T., Ottewill, R.H., and Tadros, Th.F., *J. Colloid Interface Sci.*, **100**, 332, 1984.
208. Cazabat, A.M., and Langevin, D., *J. Chem. Phys.*, **74**, 3148, 1981.
209. Tadros, Th.F., *Aspects of Appl. Biol.*, **14**, 1, 1987.
210. Guye, P., and Perrot, F.L., *Arch. Sc. Phys. et net.*, **11**, 225, 1901.
211. Abbonec, L., *Ann. Phys.*, **3**, 161, 1925.
212. Bikerman, J.J., *Surface Chemistry*, Academic Press, New York, 1958.
213. Garner, F.H., Miba, P., and Jensen, V.G., *Trans. Faraday Soc.*, **55**, 1607, 1616, 1927, 1959.
214. Davies, J.T., and Makepeace, R.W., *AIChE J.*, **24**, 524, 1978.
215. Adamson, A.W., *Physical Chemistry of Surfaces*, 4th Ed., Wiley-Interscience, New York, 1982.
216. de Gennes, P., *Adv. Colloid Interface Sci.*, **27**, 189, 1987.
217. Hartley, G.S., and Brunskill, R.G., *Surface Phenomena in Chemistry and Biology*, Pergamon Press, New York, 1958, pp. 214–223.
218. Brunskill, R.T., *Proceedings of the Third Weed Conference*, Association of British Manufacturers, 1956, p. 593
219. Seaman, D., *Chem. Ind.*, 1979, pp. 159–165.
220. Boize, L.M., Lee, G.W.J., and Tadros, Th.F., *Pest. Sci.*, **14**, 427, 1983.
221. Bikerman, J.J., *J. Colloid Sci.*, **5**, 349, 1950.
222. Furmidge, C.G.L., *J. Colloid Interface Sci.*, **17**, 309, 1962.
223. Frankel, Y.A.I., *Exptl. Theoret. Phys. (USSR)*, **18**, 659, 1948.
224. Rosano, H.K., *Men. Serv. Chem. et al. (Paris)*, **36**, 437, 1951.

225. Buzagh, A., and Wolfram, E., *Kolloid Z.*, **157**, 50, 1958.
226. Bikerman, J.J., *Ind. Eng. Chem. (Anal. Ed.)*, **13**, 443, 1941.
227. Ford, R.E., and Furmidge, C.G.L., *SCI Monograph 25*, 1967, p. 417.
228. Blake, T., in *Surfactants*, (Tadros, Th.F., ed.), Academic Press, London, 1984, p. 221.
229. Cassie, A.B.D., and Baxter, S., *Trans. Faraday Soc.*, **40**, 546, 1944.
230. Deryaguin, B.V., *C.R. Acad. Sci. USSR*, **51**, 361, 1946.
231. Harkins, W.D., *J. Phys. Chem.*, **5**, 135, 1937.
232. Dupre, A., *Theorie Mecanique de la Chaleur*, Paris (1869), p. 201.
233. Zisman, W.A., *Adv. Chem. Ser.*, **43**, 1964.
234. Fox, H.W., and Zisman, W.A., *J. Colloid Sci.*, **5**, 514, 1950.
235. Good, R.J., and Girifalco, L.A., *J. Phys. Chem.*, **64**, 561, 1960.
236. Fowkes, F.M., *Adv. Chem. Ser.*, **43**, 99, 1964.
237. Smolders, C.A., *Rec. Trans. Chim.*, **80**, 650, 1961.
238. Hartley, G.S., and Graham-Bryce, I.J., *Physcial Principles of Pesticide Behaviour*, Academic Press, London, Vol.I, 1980.
239. Tadros, Th.F., in *Microbial Adhesion to Surfaces*, (Berkley, R.C.W., Lynch, J.M., Melling, J., Rutter, P.R., and Vincent, B., eds.), Ellis Horwood, Chichester, 1980, Chap. 5.
240. Tadros, Th.F., in *Surfactants*, (Tadros, Th.F., ed.), Academic Press, London, 1984, p. 323.
241. Attwood, D., and Florence, A.T., *Surfactant Systems, Their Chemistry, Pharmacy and Biology*, Chapman and Hall, London, 1983.
242. Higuchi, W.I., *J. Pharm. Sci.*, **53**, 532, 1964.
243. Higuchi, W.I., *J. Pharm. Sci.*, **56**, 315, 1967.
244. Elworthy, P.H., and Lipscomb, F.J., *J. Pharm. Pharmacol.*, **20**, 923, 1968.
245. Chan, A.F., Evans, D.F., and Cussler, E.L., A.J. Chem., **22**, 1006, 1976.
246. Mysels, K.J., *Adv. Chem. Ser.*, **86**, 24, 1969.
247. Bates, T.R., Gibaldi, M., and Konig., J.L., *Nature*, **210**, 1331, 1966.
248. Merrill, R.C., and McBain, J.W., *J. Phys. Chem.*, **46**, 10, 1942.
249. Turner, D.J., and Loader, M.P.C., *Proceedings of the 12th British Weed Conference*, 1974, pp. 177–184.

Index